SUCCESSFUL STEM EDUCATION

A WORKSHOP SUMMARY

Alexandra Beatty, *Rapporteur*

Committee on Highly Successful Schools or Programs for K-12 STEM Education

Board on Science Education

Board on Testing and Assessment

Division of Behavioral and Social Sciences and Education

NATIONAL RESEARCH COUNCIL
OF THE NATIONAL ACADEMIES

THE NATIONAL ACADEMIES PRESS
Washington, D.C.
www.nap.edu

THE NATIONAL ACADEMIES PRESS 500 Fifth Street, N.W. Washington, DC 20001

NOTICE: The project that is the subject of this report was approved by the Governing Board of the National Research Council, whose members are drawn from the councils of the National Academy of Sciences, the National Academy of Engineering, and the Institute of Medicine. The members of the committee responsible for the report were chosen for their special competences and with regard for appropriate balance.

This study was supported by Grant Nos. DRL-1050545 and DRL-1063495 between the National Academy of Sciences and the National Science Foundation. Any opinions, findings, conclusions, or recommendations expressed in this publication are those of the author(s) and do not necessarily reflect the views of the organizations or agencies that provided support for the project.

International Standard Book Number-13: 978-0-309-21890-0
International Standard Book Number-10: 0-309-21890-X

Additional copies of this report are available from the National Academies Press, 500 Fifth Street, N.W., Lockbox 285, Washington, DC 20055; (800) 624-6242 or (202) 334-3313 (in the Washington metropolitan area); Internet, http://www.nap.edu.

Printed in the United States of America

Suggested citation: National Research Council. (2011). *Successful STEM Education: A Workshop Summary.* A. Beatty, Rapporteur. Committee on Highly Successful Schools or Programs for K-12 STEM Education, Board on Science Education and Board on Testing and Assessment. Division of Behavioral and Social Sciences and Education. Washington, DC: The National Academies Press.

THE NATIONAL ACADEMIES

Advisers to the Nation on Science, Engineering, and Medicine

The **National Academy of Sciences** is a private, nonprofit, self-perpetuating society of distinguished scholars engaged in scientific and engineering research, dedicated to the furtherance of science and technology and to their use for the general welfare. Upon the authority of the charter granted to it by the Congress in 1863, the Academy has a mandate that requires it to advise the federal government on scientific and technical matters. Dr. Ralph J. Cicerone is president of the National Academy of Sciences.

The **National Academy of Engineering** was established in 1964, under the charter of the National Academy of Sciences, as a parallel organization of outstanding engineers. It is autonomous in its administration and in the selection of its members, sharing with the National Academy of Sciences the responsibility for advising the federal government. The National Academy of Engineering also sponsors engineering programs aimed at meeting national needs, encourages education and research, and recognizes the superior achievements of engineers. Dr. Charles M. Vest is president of the National Academy of Engineering.

The **Institute of Medicine** was established in 1970 by the National Academy of Sciences to secure the services of eminent members of appropriate professions in the examination of policy matters pertaining to the health of the public. The Institute acts under the responsibility given to the National Academy of Sciences by its congressional charter to be an adviser to the federal government and, upon its own initiative, to identify issues of medical care, research, and education. Dr. Harvey V. Fineberg is president of the Institute of Medicine.

The **National Research Council** was organized by the National Academy of Sciences in 1916 to associate the broad community of science and technology with the Academy's purposes of furthering knowledge and advising the federal government. Functioning in accordance with general policies determined by the Academy, the Council has become the principal operating agency of both the National Academy of Sciences and the National Academy of Engineering in providing services to the government, the public, and the scientific and engineering communities. The Council is administered jointly by both Academies and the Institute of Medicine. Dr. Ralph J. Cicerone and Dr. Charles M. Vest are chair and vice chair, respectively, of the National Research Council.

www.national-academies.org

COMMITTEE ON HIGHLY SUCCESSFUL SCHOOLS OR PROGRAMS FOR K-12 STEM EDUCATION

ADAM GAMORAN (*Chair*), Department of Sociology and Wisconsin Center for Education Research, University of Wisconsin–Madison
JULIAN BETTS, Department of Economics, University of California, San Diego
JERRY P. GOLLUB, Natural Sciences and Physics Departments, Haverford College
GLENN "MAX" McGEE, Illinois Mathematics and Science Academy
MILBREY W. McLAUGHLIN, School of Education, Stanford University
BARBARA M. MEANS, Center for Technology in Learning, SRI International
STEVEN A. SCHNEIDER, Science, Technology, Engineering, and Mathematics Program, WestEd
JERRY D. VALADEZ, Central Valley Science Project, California State University, Fresno

MARTIN STORKSDIECK, *Director*, Board on Science Education
STUART ELLIOTT, *Director*, Board on Testing and Assessment
NATALIE NIELSEN, *Study Director*
MICHAEL FEDER, *Study Director* (until February 2011)
THOMAS KELLER, *Senior Program Officer*
REBECCA KRONE, *Program Associate*

BOARD ON SCIENCE EDUCATION
2011

BOARD ON TESTING AND ASSESSMENT

Preface

Science, technology, engineering, and mathematics (STEM) are at center stage in the education reform movement. Most people share the vision that a highly capable STEM workforce and a population that understands and supports the scientific enterprise are key to the future place of the United States in global economics and politics and to the well-being of the nation. Many schools around the country are producing students who are eager to go on to advanced study and who excel in college and in STEM careers. Many students are left behind, however. Talented and potentially eager students who may not have access to elite schools or excellent programs may never recognize their potential to excel. And too many U.S. students progress through K-12 education without attaining basic mathematics and scientific knowledge and skills, as the nation's disappointing results on international comparisons have repeatedly demonstrated.

Although all too much is known about why schools may not succeed, it is far less clear what makes STEM education effective. The Committee on Highly Successful Schools or Programs for K-12 STEM Education was created, with the support of the National Science Foundation, to explore what makes STEM education work—the schools, the practices that excellent schools may share, and conditions that enable schools to be effective. Earlier this year we issued a short report on the findings and conclusions from our work (National Research Council, 2011b). This report describes in detail what was presented and discussed at our May 2011 workshop. For that workshop, the committee's role was limited to planning: this summary has been prepared by a rapporteur, with staff assistance as appropriate. The workshop was not designed to generate consensus conclusions or recommendations but focused instead on the identification of ideas, themes, and considerations that contribute to understanding the topic; the report does not represent either findings or recommendations that can be attributed to the committee. This document summarizes the views expressed by workshop participants, and the committee was responsible only for the quality of the agenda and the selection of participants.

This workshop summary has been reviewed in draft form by individuals chosen for their diverse perspectives and technical expertise, in accordance with procedures approved by the Report Review Committee of the National Research Council (NRC). The purpose of this independent review is to provide candid and critical comments that will assist the institution in making its published report as sound as possible and to ensure that the report meets institutional standards for objectivity, evidence, and responsiveness to the charge. The review comments and draft manuscript remain confidential to protect the integrity of the process.

We thank the following individuals for their review of this report: Mark Berends, Department of Sociology, Director of the Center for Research on Educational Opportunity and the National Center on School Choice, University of Notre Dame; Theodore R. (Ted) Britton, Associate Director, National Center for Improving Science Education, WestEd; Patti Curtis, Managing Director, Washington Office, Museum of Science, Boston, and National Center for Technological Literacy; Jacob Foster, Science, Technology, and Engineering, Massachusetts Department of Elementary and Secondary Education; Joseph Krajcik, Science Education, Codirector, IDEA Institute, School of Education, University of Michigan; Christopher C. Lazzaro, Director of Science Education, Research and Development, College Board; and Walter G. Secada, Professor and Senior Associate Dean, School of Education, University of Miami.

Although the reviewers listed above provided many constructive comments and suggestions, they were not asked to endorse the content of the report nor did they see the final draft of the report before its release. The review of this report was overseen by Carlo Parravano, Merck Institute for Science Education. Appointed by the NRC, he was responsible for making certain that an independent examination of this report was carried out in accordance with institutional procedures and that all review comments were carefully considered. Responsibility for the final content of this report rests entirely with the committee and the institution.

Adam Gamoran, *Chair*
Committee on Highly Successful Schools or Programs for K-12 STEM Education

Contents

APPENDIXES

1

Introduction

The phrase "STEM education" is shorthand for an enterprise that is as complicated as it is important. What students learn about the science disciplines, technology, engineering, and mathematics during their K-12 schooling shapes their intellectual development, opportunities for future study and work, and choices of career, as well as their capacity to make informed decisions about political and civic issues and about their own lives. A wide array of public and personal issues—from global warming to medical treatment to social networking to home mortgages—involves science, technology, engineering, and mathematics (STEM). Indeed, the solutions to some of the most daunting problems facing the nation will require not only the expertise of top STEM professionals but also the wisdom and understanding of its citizens.

Education in the STEM areas takes many forms in the United States. Though there are compelling reasons for concern about the quality and effectiveness of the education many students receive in these disciplines, there are also many clear success stories. Policy makers and others have looked for ways to identify the schools and approaches that are most successful—and the characteristics that account for their success—so that their models for best practice can be replicated.

At the request of the office of U.S. Representative Frank Wolf (R-VA), the National Science Foundation asked the National Research Council to explore these issues, and, under the auspices of the Board on Science Education and the Board on Testing and Assessment, the Committee on Highly Successful Schools or Programs for K-12 STEM Education was

BOX 1-1
Charge to the Committee

An ad hoc steering committee will plan and conduct a public workshop to explore criteria for identifying highly successful K-12 schools and programs in the area of STEM education through examination of a select set of examples. The committee will determine some initial criteria for nominating successful schools to be considered at the workshop. The examples included in the workshop must have been studied in enough detail to provide evidence to support claims of success. Discussions at the workshop will focus on refining criteria for success, exploring models of "best practice," and an analysis of factors that evidence indicates lead to success. The discussion from the workshop will be synthesized in an individually authored workshop summary.

formed to carry out this work. The committee was charged with "outlining criteria for identifying effective STEM schools and programs and identifying which of those criteria could be addressed with available data and research, and those where further work is needed to develop appropriate data sources." The detailed charge is shown in Box 1-1.

To carry out part of its charge, the committee organized a workshop, held in May 2011, that had three goals:

1. describing the primary types of K-12 schools and programs that can support successful education in the STEM disciplines;
2. examining data and research that demonstrate the effectiveness of these school types; and
3. summarizing research that helps to identify both the elements that make such programs effective and what is needed to implement these elements.

This report is a summary of that workshop.[1] The remainder of this chapter elaborates on why STEM education is so important and on the complexity of the task of identifying the features that are essential to successful outcomes for students. Chapter 2 explores four different basic

[1]The workshop sessions included formal presentations and structured panel discussions. Because the primary purpose was for the committee to support its development of consensus findings for a separate report, there were opportunities for the committee to question presenters. There were also some opportunities for general discussion. This summary synthesizes the material presented and highlights from the questions and discussion. The committee also wrote a report summarizing its findings, conclusions, and recommendations on STEM education (National Research Council, 2011).

types of schools that deliver STEM education in the United States, looking both at research on each type and a few example schools. Chapter 3 addresses the research on practices and approaches to science and mathematics education, and Chapter 4 explores research on school conditions that support effective STEM education. The closing chapter summarizes the major points that emerged from the workshop discussion, with a focus on goals for translating the next generation of standards (both for the Common Core and the Next Generation Science Standards) into curricula, professional development programs, and assessments. Future research needs also were discussed. Following the list of references are four appendices. Appendix A provides the agendas for the workshop held May 10-12, 2011. Appendix B presents a list of registered workshop participants. Papers commissioned for the workshop are listed in Appendix C, and biographical sketches of committee members can be found in Appendix D.

THE IMPORTANCE OF STEM EDUCATION

STEM education has many potential benefits for individuals and for the nation as a whole, Norman Augustine explained in an opening presentation. One factor that sets it apart from other branches of academic study for many policy makers is that literacy in STEM subjects is important both for the personal well-being of each citizen and for the nation's competitiveness in the global economy. Various studies, Augustine explained, show that between 50 and 85 percent of growth in the U.S. gross domestic product over the past 50 years was accounted for by advancements in science and engineering. He also noted that the U.S. Commission on National Security, which issued its report early in 2001, highlighted the two greatest threats facing the country as terrorism on U.S. soil and "the failure to properly manage our educational system and our investments in research."

Rising Above the Gathering Storm (National Academy of Sciences, National Academy of Engineering, and Institute of Medicine, 2007), which reviewed the factors that influence U.S. competitiveness, highlighted the critical importance of STEM education in its recommendations. Drawing on a recent update of that report (National Academy of Sciences, National Academy of Engineering, and Institute of Medicine, 2010), Augustine described a few of the reasons why the United States needs to improve STEM education. "We like to think of America as being first in everything," he noted. But, for example, the United States ranks 6th among developed nations in innovation-based competitiveness, 11th in percentage of young adults who have graduated from high school, 15th in science literacy among top students, and 28th in mathematics literacy among top students.

On the basis of these data and other evidence of ways the United States is falling short in international comparisons, the National Academy of Sciences, National Academy of Engineering, and Institute of Medicine (2007) recommended a focus on improving STEM education; it highlighted parental interest and support and qualified, engaged teachers as the essential ingredients. In the 5 years between the report and the updated volume (National Academy of Sciences, National Academy of Engineering, and Institute of Medicine, 2010), Augustine added, 6 million more U.S. young people dropped out of school while many other nations continued to improve their STEM education. At the same time, the U.S. higher education system—widely regarded as first in the world—is under threat, he added. Severe budget cuts at state universities and losses in endowment funds at private ones have meant loss of faculty and other resources. Universities in other countries, he noted, "have lists of faculty members" they want to recruit, and some are recruiting promising high school students as well.

As the U.S. population changes in composition, "we are going to fall further and further behind," Augustine argued, if schools are not able to engage students from groups that have traditionally been underserved in STEM education. "We need more pathways for top students to really excel . . . and we need more alternate pathways for the kids who don't want to become scientists, but still need to be science-literate" he concluded.

DEFINING SUCCESS

The committee was asked to identify schools that have been highly successful at K-12 STEM education and to draw lessons for schools across the country, committee chair Adam Gamoran explained, but he stressed that this is a more complex challenge than it might seem. STEM encompasses many disciplines and kinds of education, and there are many ways to define it. Because of limits to the time and resources available for this project, the committee focused on mathematics and science. The bulk of the research and data concerning STEM education at the K-12 level relates to mathematics and science education. Research in technology and engineering education is less mature because those subjects are not as commonly taught in a K-12 context, but the committee fully recognizes the importance of engineering and technology education, of conceptual connections among STEM subjects,[2] and of other stages and types of schooling (including informal STEM learning).

---

[2]The nature and potential value of integrated K-12 STEM education is the focus of another, ongoing study of the National Academy of Engineering and the National Research Council by the Committee on Integrated STEM Education. It is expected to be completed in 2013.

Gamoran and committee member Barbara Means outlined three key questions and issues the committee considered in designing the workshop: What is "success"? How is it judged? What are the elements of success?

Successful at What? In general, the STEM schools that graduate the largest numbers of successful STEM students are the ones in which the largest numbers of well-prepared and highly motivated 9th graders enroll, Means observed. Thus, it is necessary to disentangle the effects of the schools from the effects of student selection; a more precise question is which schools and programs add the most value for the students they serve.

How Should Success Be Judged? There are several valued outcomes of STEM education. Goals include preparing top students for advanced degrees and technical careers in STEM fields, developing science literacy for all students, helping all students prepare for college, and equipping the future workforce to prosper individually and support the nation's prosperity. In identifying successful schools and programs it is necessary to understand their goals and the students they serve. For example, Means noted, the Georgetown Center on Education and the Workforce has estimated that in 2018, just 24 percent of the STEM-related jobs are going to require a graduate degree, 44 percent will require a bachelor's degree, and 20 percent will require either an associate's degree or a certificate for some other postsecondary program. All of the workforce options will be important to the country's economic competitiveness.

Any broad look at the effectiveness of STEM education in the United States, Means added, must also take into account the changing demographics of the population at large. The fastest growing group is low-income Hispanics, and this group also has among the lowest rates of participation in STEM occupations. Thus, success could be judged not only on how many successful graduates are produced, or average achievement, but also on how effectively the achievement gaps between different groups are narrowed.

What Elements Make Schools or Programs Successful? Programs vary not only in their goals and in the students they serve, but also in the geographic, educational, and demographic contexts in which they are located, among other factors. They operate within an education system that has many interacting layers—a "complex ecology," as Means phrased it—so a program or practice that works well with a particular group of students in one context may not work well with another. Programs and schools themselves are complex, and specific features or ways of implementing a given program in a given school may be very important to their success but difficult to isolate.

To help answer these questions, the committee developed a frame-

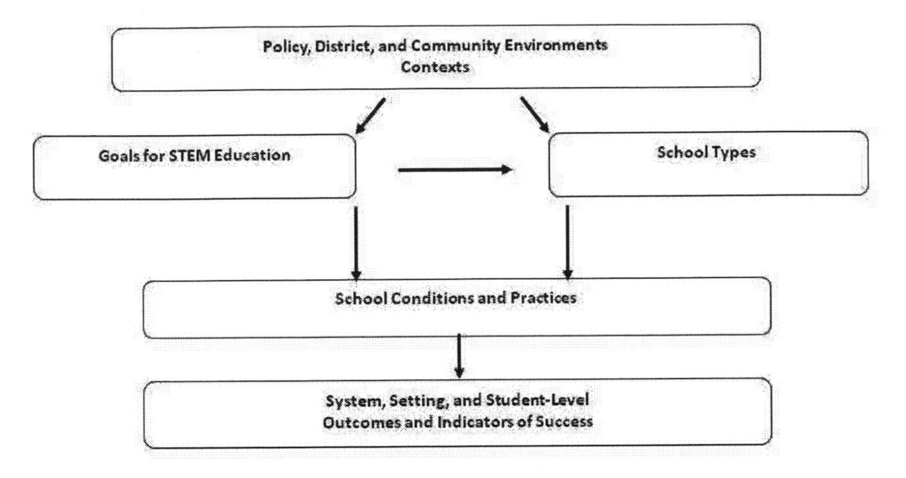

FIGURE 1-1 A framework for understanding what constitutes success in K-12 STEM education.

work for understanding what constitutes success in K-12 STEM education, which guided the planning for the workshop. This framework, shown in Figure 1-1, depicts the factors that influence the effectiveness of STEM education. For example, the context in which education takes place—the upper-most box—determines the curriculum, the resources, the priorities, and students' expectations and motivation. The program's specific goals, such as preparing top students for advanced study and challenging careers, reducing achievement gaps, and/or improving math and science literacy for all students, for example, would then dictate the standards by which the program is judged. Schools and programs have very different structures and these also must be taken into account, as must specific conditions and practices within programs. Measuring success also entails identifying specific indicators of desired outcomes. Test scores are frequently used, but course taking, college readiness and performance, choice of major, and choices and performance in the workforce are some of the other outcomes that must be considered. The workshop sessions explored these points and the available research.

Means noted that with only about 40 percent of students leaving high school prepared for college-level mathematics, "we need to do a much, much better job with many more students." What is needed is a system that is highly effective for each of the purposes and goals of STEM education, and effective for different students in different contexts. There are no easy answers, she added, and in many cases there is no solid evidence at all about best practices. The workshop presenters were asked to highlight both what is and is not known, to frame the problem, and to help identify the next steps for the research that is needed to answer the questions.

2

Four Kinds of Schools

Some kind of STEM education is offered in virtually every school, but the committee identified four broad categories of programs that offer a special emphasis on these subjects (see National Research Council, 2011):[1]

- Elite or selective STEM-focused schools. These schools serve only highly motivated and able students and focus on preparing them for ambitious postsecondary study and STEM careers.
- Inclusive STEM-focused schools. These schools do not have admissions requirements but offer specialization in one or more of the STEM disciplines. Many have the mission of helping students from population subgroups who are not well represented in STEM fields prepare for college study and STEM careers.
- STEM-focused career and technical education (CTE) schools or programs. CTE education may be offered in high schools that make this a theme, in such programs as career academies within comprehensive high schools, or in regional centers that serve many schools (Stone, 2011). Such programs are designed to prepare students for a broad range of STEM careers and often focus on engaging students at risk for dropping out of school.

[1]The workshop focused most on mathematics and science education, in part because there is more research and data for these two areas than for technology and engineering education.

- STEM programs in comprehensive schools that are not STEM focused. The majority of the nation's schools are comprehensive, and thus they educate many of the students who go on to STEM careers. Many of these schools offer advanced coursework through the Advanced Placement and International Baccalaureate Programs and other opportunities for highly motivated students.

Presenters reviewed research and perspectives on each of these school types.

SELECTIVE SCHOOLS

Focusing on the students who are the most interested and able may be the best known way to emphasize STEM education in school—but even in the category of schools with selective admissions criteria there are many approaches.

Example: A Residential School in a High-Tech Region

The North Carolina School of Science and Mathematics was founded in 1980, and this residential school was the first of its kind, Todd Roberts explained.[2] It serves 680 students in the 11th and 12th grades from every North Carolina district, and it also offers distance learning opportunities to an additional 800 students across the state. Admissions considerations include SAT scores, grades, and ambitious course taking. The school's curriculum provides a special focus in mathematics, science, and technology, along with a full complement of academic study. Though not a part of the state's public K-12 system, it is supported by the state and charges no fees to students. Since 2007 the school has been a constituent member of the University of North Carolina System. More than 7,000 students have graduated from the program to date, Roberts noted, and 60 percent have gone on to college study and careers in STEM fields.

A principal benefit of the program, in Roberts' view, is that it provides students from every part of the state with the opportunity to pursue advanced learning opportunities and to do so with a group of students who are equally excited about science and mathematics. In response to a question, Roberts noted that the school has a program for identifying students before high school who might be interested in attending and preparing them either for applying to the school or succeeding elsewhere. The program promotes collaboration among the students—there is no class rank—and encourages all students to pursue opportunities to

[2]For more information about the school, see http://www.ncssm.edu/ [June 2011].

conduct research and work with mentors. Because the school is located in the Research Triangle Park area of North Carolina, there are numerous universities and research facilities close by, and the students benefit from these resources both during the school year and through summer internships.

Of the school's graduates, 63 percent return to live and work in North Carolina after college, Roberts added. The state leaders who established the school through legislation had envisioned that it would not only serve as a model for educational improvement, but also support the state's economic goals by providing a steady supply of highly qualified workers. From the state's perspective, establishing a specialized school focused on science and mathematics that would be independent of the school system has paid off.

Graduates of Selective Specialized Schools: Research Findings

Looking beyond a single school, Robert Tai and Rena Subotnik described preliminary findings from a study they are conducting of graduates from selective public high schools of science, mathematics, or technology (Subotnik, Tai, and Almarode, 2011). The study is designed to assess the value these schools add by developing and maintaining the supply of students who pursue advanced degrees and careers in STEM fields. The researchers have surveyed students 4-6 years after graduation and combined the results with other data available about the cohort from the National Education Longitudinal Study (NELS)[3] to develop answers to two questions: Are these graduates more likely to enter STEM programs in college and STEM careers than other students? Which educational models used in their schools seem to yield the most students who pursue STEM-related study and careers?

First, Tai noted, there is no clear definition of this type of school. For their study, they identified four subtypes among the selective schools that specialize in STEM education: residential programs; comprehensive programs that have a special focus on STEM; specialized STEM programs that operate within a larger school; and half-day programs, in which students commute between a specialized program and their home schools. Finding that these schools offer very different experiences for students, Tai and Subotnik collected data from two of each of these four types. Although there is variation among the subtypes, some common features include advanced STEM coursework, expert teachers, like-minded peers who are interested in STEM, and opportunities for independent research. Tai and Subotnik's primary outcome measure was whether or not the

[3]For more information about NELS, see http://nces.ed.gov/surveys/nels88/ [June 2011].

TABLE 2-1 Students' Goals and Choices of Major by High School Type (in percentage)

High School Type	Entered High School Intending to Pursue STEM Career	Chose STEM-Related College Major
Residential specialized	77.9	56.8
Comprehensive specialized	59.8	42.2
Specialized school within a school	78.9	65.5
Half-day specialized	74.5	50.9

SOURCE: Adapted from Subotnik, Tai, and Almarode (2011).

students reported having completed an undergraduate major in a STEM field, and they asked a range of questions about the student's high school experiences.

Tai set the context by reporting on NELS data about students who entered high school thinking they were interested in science and remained engaged in science by the end of their college careers. Among all students who began high school interested in science, 40.7 percent completed an undergraduate degree in science; and among those who were interested in science and also were high performers in science and mathematics,[4] 46.6 did so. Tai and Subotnik's data show that students who entered high school interested in science and also attended a specialized high school program are significantly more likely to stay in science—64.9 percent of them did so. For comparison, students who were *not* initially interested in science but switched into a science field are much less likely to choose an undergraduate science major: 21.9 percent of all students were in this group; 34.0 percent of the high performers were; and 27.5 percent of those who attended a specialized high school but were not initially interested in science were. In other words, students who are interested in science prior to high school are significantly more likely to stay in the field.

There was also variation both in students' goals as they entered high school and in their ultimate choices of major across the four types of specialized schools: see Table 2-1.

Tai and Subotnik used statistical procedures to determine how much of this variation could be accounted for by differences among these school types and how much could be accounted for by variations among the students. They calculated that school-level differences accounted for 3.6

[4]Tai explained that their comparison group was identified through a national academic talent search program and was composed of students, matched by age, grade, and standardized test scores, who had also chosen to participate in formal science and mathematics activities.

percent of the variation in whether or not students completed an undergraduate science major: thus, 96.4 percent was accounted for by student differences.

Additional survey questions allowed them to explore some of the differences in the students' experiences. Their preliminary data indicate that, among graduates of specialized STEM high school programs:

- Students who participated in or conducted original scientific research while in high school were 70 percent more likely to major in a STEM field than those who did not.
- Students who participated in internships or had mentors were 20 percent more likely to major in a STEM field than those who did not.
- Students who reported a strong sense that they "belonged" during their high school years were 22 percent more likely to choose a STEM major than those who did not report "belonging."
- Students who reported that their teachers frequently made connections across the curriculum were 23 percent more likely to choose a STEM major than those who did not so report.

Each individual factor, Tai observed, may not have a profound effect on its own, but taken together "they open up a pathway" for students into STEM fields. These preliminary data provide a more detailed picture of why students who graduate from specialized schools pursue STEM fields in college at a rate nearly 50 percent higher than that of other students.

INCLUSIVE STEM-FOCUSED SCHOOLS

Schools and programs that offer a broader population of students the chance to focus on STEM subjects have some things in common with the selective schools, but there are differences as well.

Example: A Hybrid School

Montgomery Blair High School, located in a Washington, DC, suburb, offers some of the features of both types.[5] This public school, which serves a demographically diverse population, is home to a highly selective STEM magnet program. Principal Daryl Williams explained that it is part of a Montgomery County network of programs located within neighborhood schools but designed to attract students from a wider geographic area

[5]For more information about Montgomery Blair High School, see http://www.montgomeryschoolsmd.org/schoolodex/schooloverview.aspx?s=04757 [June 2011].

by offering academically demanding programs. Montgomery Blair offers all students the chance to study in one of five academies: entrepreneurship and business management; human service professions; international studies and law; media literacy; and science, technology, engineering, and mathematics. The school also has two magnet programs—one in communication arts and one in science, mathematics, and computer science. Williams noted that 400 of the school's 2,864 students are enrolled in the science and mathematics magnet program, which is distinct from the five academies (and thus travel by bus from neighborhoods outside the school's catchment area).

The science, technology, engineering, and mathematics academy and the related magnet program share the goals of giving students the opportunity to pursue independent and collaborative research projects, as well as to work with mentors at local businesses and research organizations.

A Texas STEM Program: Research Findings

In 2003, Texas inaugurated a public-private partnership program, the Texas High School Project (THSP), dedicated to helping low-income students prepare for postsecondary study and helping low-performing schools improve. The Texas Science, Technology, Engineering, and Mathematics Initiative (T-STEM) is one element of that initiative, Viki Young explained (see Young, 2011). Since 2006 the state has invested $120 million to open 51 high school academies and 7 technical assistance centers that provide professional development and other services to Texas schools. A key goal for these centers is to improve outcomes for all schools, not just the academies, which are designed as demonstration schools. The academies do not have selection requirements—students are admitted by lottery if the school is oversubscribed. Because T-STEM is intended to serve high-need students, the academies are located in high-need areas and are required to maintain student populations in which more than 50 percent of the students are economically disadvantaged or members of traditionally disadvantaged ethnic and racial groups.

Young and her colleagues used data from a 4-year longitudinal evaluation of the THSP to analyze the effects of this program on student outcomes (Young, 2011). They used both qualitative and quantitative methods to study the implementation of T-STEM. The variety of outcome measures used to gauge T-STEM's influence included results from the Texas Assessment of Knowledge and Skills (TAKS) in several subjects, passage of Algebra I by 9th grade, grade promotion, and rates of absenteeism.

The preliminary results, Young explained, indicate that students who attended the T-STEM academies performed slightly better than their peers at comparable schools[6] in both mathematics (9th and 10th grades) and science (10th grade; there is no 9th grade science test). The T-STEM students were more likely than their peers to pass all of the required parts of the TAKS, and T-STEM 9th graders have lower rates of absenteeism.

Young cited several factors that may have influenced these outcomes. First, both students and faculty come to the T-STEM academies by choice. Though families may not have sought out a STEM focus, they have sought an academically rigorous program and are likely to be more academically motivated than other families. Student attrition may also affect the results. The academies report that students who find the workload too great or do not feel that they fit in tend to leave: 22 percent of students leave between 9th and 10th grade and 35 percent leave between 10th and 11th grade. These "dropouts" are important because TAKS results are reported only for students who had been at their schools since 9th grade.

The academies also offer a number of supports for students who may not be well prepared for a rigorous STEM curriculum when they enter. The supports include one-on-one tutoring, extra instruction for small groups, and credit recovery (opportunities to retake a course in which a student was not successful). Although such supports are also found at other schools, Young highlighted the "climate of high expectations" at the T-STEM academies, the opportunities for close relationships between students and faculty that result from the time set aside for advisory groups and regular check-ins, and the supports for college preparation activities. The academies are small (100 students per grade), and Young pointed out that this allows all students to have teachers who know them as individuals and also allows teachers to track students' progress. However, she noted, the T-STEM academies are not uniformly implementing the blueprint that was intended to guide them.

The T-STEM academies strive for other outcomes, such as college readiness, mastery of 21st century skills, and involvement in out-of-school experiences that prepare them for STEM careers. However, these sorts of outcomes have not been consistently measured, in part because the T-STEM program has only been in place for a few years. It will take time before these kinds of outcomes for T-STEM students develop and can be measured, though Young suggested that they may be the most significant. Over time, she suggested, it will be important to study the math and science literacy of T-STEM students, their readiness for college, and the rate at which they choose to major in STEM fields. In addition, she believes, researchers should study the effects of inclusive STEM schools in other states, and should build the capacity to look longitudinally at high school

[6]The researchers used statistical procedures to identify comparison schools that were similar to the T-STEM academies: see Appendix A in Young (2011).

and postsecondary experiences. She also noted that they should seek ways to control for the selection bias that may have affected the current results and look more closely at the specific features of the approach used at the T-STEM academies to identify those most closely associated with desired outcomes.

STEM-FOCUSED CAREER AND TECHNICAL EDUCATION

Defining CTE—and understanding its relationship to STEM education more broadly—is no less complicated than defining the other categories of STEM education. Nevertheless, James Stone pointed out, the primary goal for CTE is to develop technologically proficient workers.

Example: Many Options in a Single School

Lake Travis High School, a school of just over 2,000 students in Austin, Texas, has organized its curriculum into six institutes: advanced science and medicine; mathematics, engineering, and architecture; humanities, technology, and communications; veterinary and agricultural science; business, finance, and marketing; and fine arts. As Jill Siler explained, the district has just one high school and as the population has grown, it sought a way to provide students with a small-school experience without building a second high school.

The institutes are designed to be flexible—students select their course of study and can move between the institutes. The school is run on an alternating block schedule, which allows time for longer class periods. Many of the credits are articulated so students can earn credits at the local community college, and the math, engineering, and architecture institute offers six year-long engineering courses through Project Lead the Way.[7] In the STEM-related institutes, students can further specialize and can also undertake field work or find mentors at local research or other sites or engage in distance learning.

Types of Career and Technical Education

Lake Travis High School's flexible approach to providing career and technical education—in which students can partake of as much of it as they wish—can be found in many models. As Stone explained, more than 90 percent of high school students take at least one CTE course, though only 17 percent do so as part of CTE focus or concentration (Levesque

[7]Project Lead the Way provides STEM curricula to middle and high schools. For more information, see http://www.pltw.org/ [July 2011].

et al., 2008). While the goals for career and technical education are not precisely the same as those for STEM education, he added, all career and technical education is related to some aspect of the STEM fields, and he sought to identify which CTE approaches most effectively promote the learning of STEM subjects (Stone, 2011).

Stone identified five structures through which career and technical education is generally offered, though they overlap in some cases. Two are entities focused completely on CTE: regional career technical centers and CTE high schools. Three other approaches are generally housed in traditional comprehensive high schools: career academies, programs of study, and career clusters or pathways.

Regional Career Technical Centers

Regional career technical centers are designed to provide 11th and 12th grade students with instruction not available at their home schools, and students typically spend half days in the centers, although a few are full-day. Stone said that there is limited evidence about the effectiveness of these programs, in part because student data are collected by the home schools and cannot easily be linked to time spent in regional centers. He noted that many center faculty lack traditional academic credentials because the focus is on preparation for occupations and instructors need to be skilled in the occupation, preparation for which comes through non-college providers (e.g., apprenticeship, work experience), and the centers often have limited academic offerings. There are approximately 1,200 such centers in the United States.

CTE High Schools

CTE high schools offer core academic coursework while also requiring students to complete CTE courses in order to graduate. Students are asked to choose a career focus, usually at the beginning of 9th grade. There are approximately 900 such schools in the United States. One such school is Blackstone Valley Technical High School in Massachusetts, a school in which students perform above state averages on the Massachusetts Comprehensive Assessment System and also have a graduation rate that is 15 points above the state average. Students must complete 32 credits of vocational/technical education classes, choosing from options that include auto body and auto tech, carpentry, culinary arts, and health services, as well as more STEM-intensive areas, such as electronics and information technology. Students may also take Project Lead the Way courses. However, Stone noted that the school is selective and that the percentages of low-income and minority students in the school's population are lower than state averages. Some data on these programs are available in the Common Core of Data collected by the National Center for Education Statistics.

Career Academies

Career academies allow students to organize their studies around a career theme, such as health, computer technology, or business and finance; to build relationships with faculty devoted to that theme; and to be part of a group of students at their home school who share their interests. Such programs have become very common, Stone observed; approximately 2,500 high schools now have them.

Programs of Study

"Program of study" is a term used in the federal Carl D. Perkins Career and Technical Education Improvement Act of 2006 to describe programs that help students make the transition from secondary to post-secondary schooling. State and local agencies that receive federal funding through this legislation are required to offer programs that coordinate academic and CTE coursework and prepare students to obtain industry or academic credentials.[8]

Career Clusters or Pathways

Career clusters and pathways describe ways of grouping coursework related to different occupations or industries to help guide students in choosing a sequence of high school courses that will prepare them for a field in which they are interested. Sixteen clusters have been defined by the states' "Career Clusters Initiative."[9] One is science, technology, engineering, and mathematics, but a number of the others (e.g., agriculture, information technology, manufacturing) relate to STEM education more broadly defined.

Approaches to Career and Technical Education

Regardless of the school structure, Stone explained, there are a range of curricula and pedagogical approaches to career and technical education. For example, Project Lead the Way is a very well-known pre-

---

[8]For more information, see http://cte.ed.gov/nationalinitiatives/localstudyimplementation.cfm [August 2011].

[9]For more information, see http://www.careerclusters.org/ [August 2011].

engineering curriculum that schools can adopt. It focuses on providing hands-on experiences that prepare students for engineering-related careers. To date there has been one independent longitudinal study of this program and its outcomes, by Schenk and colleagues (Schenk et al., 2009). They found that students who participate in Project Lead the Way are more likely than their peers to be enrolled in a gifted and talented program, have better math and science skills prior to enrolling, and perform better on state assessments. They also are less likely than their peers to be eligible for free and reduced-price lunch, to be female, and to belong to a minority group. The program's own research shows that it is effective at reducing achievement gaps among student groups and improving both test scores and college readiness.

Other approaches include curriculum integration, in which links among academic disciplines are explored and students have opportunities to learn about the real-world applications of mathematics and science; project-based learning, in which students conduct extended inquiry projects; and work-based learning, in which supervised learning activities take place at a work site.

Stone described a study he and colleagues conducted to determine whether enhancing the mathematics instruction embedded in a technical education program would build students' mathematics skills while still developing the intended technical skills (Stone et al., 2008). In this study of 200 teachers and 3,000 students, teachers were randomly assigned to either the experimental or control situation. The study included programs in agriculture, information technology, automotive technology, health, and business, but the focus was the mathematics instruction (in applied, traditional, and college preparatory mathematics) that occurred naturally as part of the curriculum in each area. The researchers were exploring a model of curriculum integration and professional development called Math-in-CTE and were careful to monitor the fidelity with which the teachers implemented the approach.

The results showed that students in the experimental classes scored significantly higher than those in the control classes on both the Terra Nova and Accuplacer mathematics assessments, without any loss in the development of occupational or technical skills. Work is currently under way to explore the effects of a similar model for enhancing science instruction in a CTE context.

Stone suggested that other pedagogical approaches, such as project-based learning and work-based learning, also hold promise as means of enhancing STEM learning, but there is as yet limited evidence for these approaches. There is also very little evidence regarding the effectiveness of the different structures through which career and technical education is delivered. Stone noted that it can be difficult to distinguish STEM education from CTE approaches for purposes of research, but he suggested that there are opportunities to address important questions with rigorous research. In his view, further exploration of ways to improve science and mathematics instruction in the context of career and technical education, and of how conducive a variety of CTE approaches are to efforts to boost science and mathematics, would be very useful. He noted that spending more time in science and mathematics classes is not likely to be as beneficial as would finding better ways to use already available instructional time to build important skills.

STEM EDUCATION IN NON-STEM-FOCUSED SCHOOLS

The majority of U.S. students are educated in traditional schools, and many of those schools do an excellent job at STEM education. Many high schools offer advanced placement and international baccalaureate courses for highly motivated students. Many STEM-related programs are available to middle and high schools, and some schools excel even without special programs. Several participants discussed different schools and their approaches to STEM education.

Example: A Diverse K-8 School

Janet Elder, the principal of Christa McAuliffe School in Jersey City, New Jersey, said that no one factor is responsible for what the school has achieved. The school serves a very diverse population with a high mobility rate: of its 1,000 students, 82 percent are eligible for free and reduced-price lunches, and 65 percent speak a language other than English at home. Nevertheless, in 2010, 90 percent of the school's 8th graders and 91 percent of 4th graders scored at the proficient level or above on New Jersey's science assessment. The school has won awards in science: most notably, it was a 2010 finalist in the INTEL School of Distinction competition and the 2011 state winner of the Disney Planet Challenge, and it has won other awards and grants.

The school offers a challenging standards-based curriculum for all students, Elder explained, as well as a number of special programs, including after-school tutoring, science and technology classes, and robotics. Among 8th graders, 25 percent take both algebra and physics, and, by district policy, the other 75 percent are tracked into the general 8th grade curriculum. "That is not by my choice," Elder stressed. She is hoping to significantly increase participation in the challenging courses and to offer teachers professional development so that they can become certified to teach the high school level material, but she has not yet received approval from the state superintendent for these proposals.

Elder attributes the school's success to consistency in three areas: building community involvement, through a range of parent resource and outreach activities, including technology classes; student engagement, which is developed through a large number of in-class and extracurricular opportunities that target students' interests; and instructional leadership, fostered through professional development, peer coaching, and opportunities to collaborate. She stressed that strong teachers have been critical to the school's success. Yet, she noted, other factors have impeded the school's progress. High student mobility is perhaps their greatest challenge, and it is exacerbated by state testing requirements that drain time and resources. She worries that the state's assessments will not soon be aligned with the Common Core standards, which New Jersey has adopted: "We are going to be teaching something that isn't going to be tested and we will be a failing school in a few years."

Effective Mathematics Education

Many individual schools are very effective, William Schmidt agreed, but, on average, U.S. students are not excelling in mathematics and science and even the most elite U.S. students do not compare well with their international counterparts (Schmidt, 2011). Mathematics scores on the National Assessment of Educational Progress have improved since the mid-1990s, he noted, but three-quarters of 8th graders still enter high school not having reached the proficient level and three-quarters of high school students graduate with "a relatively poor grasp of mathematics." Even the most elite U.S. students were last in physics and close to the bottom in mathematics in a comparison with their counterparts in other nations on the Trends in International Mathematics and Science Study.

Based on his own and other research, Schmidt has identified five elements he views as essential to reforming mathematics education: (1) curriculum, (2) teacher knowledge, (3) public support for demanding standards and requirements, (4) student engagement in STEM areas, and (5) instructional leadership.

His focus is curriculum, and Schmidt observed that it is important to consider not only the curriculum that a school system intends to present, but also the content that is actually delivered by teachers. In looking at a school's curricula, one must ask how coherent it is in the way it structures the material to be taught in each grade; what its degree of focus is, in terms of how much exposure students actually have to different topics and how many are presented at each grade; and how rigorous (cognitively complex) it is. In each of these areas, curricula in the United States leave much to be desired, in his view.

Other countries tend to have more rigorous curricula, Schmidt explained. In U.S. middle schools, for example, "we are teaching arithmetic and what I call rocks and body parts, whereas in the rest of the world they are teaching chemistry, physics, algebra, and geometry. They teach their children how the brain sees as the photons enter the eye producing a biochemical reaction. We teach the parts of the eye."

STEM disciplines have a logical structure, he added. Mathematics is very hierarchical, with concepts that build cumulatively. Knowledge in the science disciplines is less hierarchical, but there is still a logical structure that defines the bodies of knowledge. That structure should guide the mapping of topics for school curricula, he observed, so that students can connect the deeper principles. The countries whose students perform at the highest levels tend to have curricula that are very coherent and focused—that is, they cover a few key topics at each grade from K through 8 and progress in a logical fashion from the most basic concepts to more complex material: see Table 2-2. Curricula in the United States, generally set at the state level, are far less orderly, Schmidt said, as Table 2-3 shows. (This table is a graphic representation of the material covered by these curricula. Because it is large, it is printed at a scale that illustrates patterns in the lack of consistency on the coverage but does not allow readers to discern the text.) However, the Common Core mathematics standards more closely resemble the pattern for the high-performing countries: see Table 2-4.

Table 2-2 also suggests the rigor of the curricula used by top-performing countries with all students, not just those who are already beginning to excel in STEM subjects. By 8th grade, for example, students are learning about congruence, the rational number system, the field theorems, and slope trigonometry. In Schmidt's view, U.S. non-STEM schools have an obligation to provide equal opportunities for all children: "If there are three 2nd grade classrooms, they all should be covering the same basic content. We shouldn't be trying to differentiate and allow teachers to make decisions about what content to cover." In other countries, he added, the teachers do not decide what material to cover: "The pedagogy is their purview," but content is determined by specialists in curriculum development. In contrast, his research shows wide variations in what is presented in classrooms at the same grade level, as well as in the amounts of time devoted to different topics at the same grade. Schmidt said he is not suggesting that classrooms should be completely uniform, but when coverage of basic arithmetic in grade 2, for example, varies from 20 days to 140 days in a year, as he has found, "You can see that there is something afoul."

Schmidt also argued that tracking of students in non-STEM schools creates problems. His research has shown that students in schools that offer only one curriculum learn significantly more mathematics than those

TABLE 2-2 Top-Achieving Countries in Mathematics, by Type of Curriculum

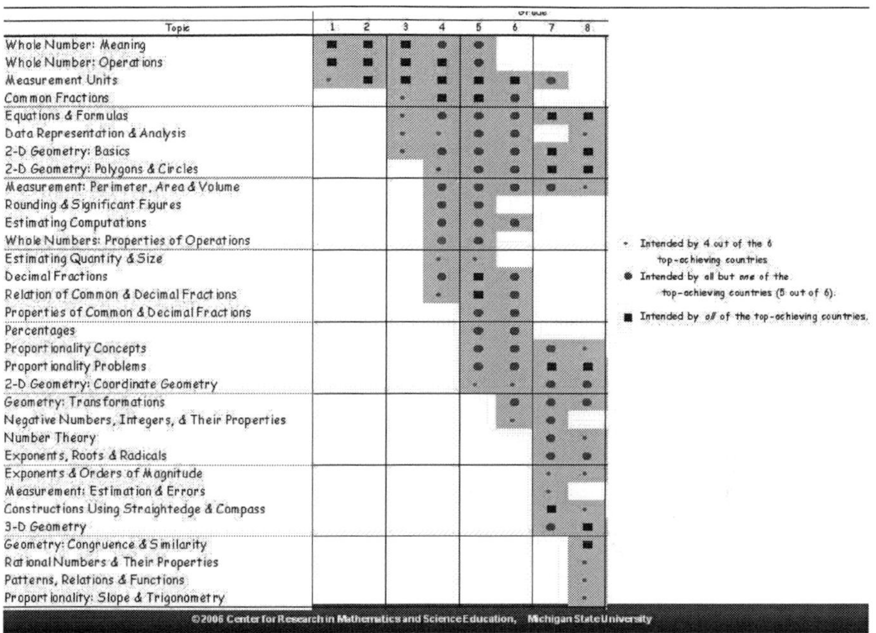

Topic	Grade 1	2	3	4	5	6	7	8
Whole Number: Meaning	■	■	■	●	●			
Whole Number: Operations	■	■	■	■	●			
Measurement Units	•	■	■	■	■	■	●	
Common Fractions			•	■	■	●		
Equations & Formulas			•	●	●	●	■	■
Data Representation & Analysis			•	•	●	●		•
2-D Geometry: Basics			•	●	●	●	■	■
2-D Geometry: Polygons & Circles				•	●	●	■	■
Measurement: Perimeter, Area & Volume				●	●	●	●	•
Rounding & Significant Figures				●	●			
Estimating Computations				●	●	●		
Whole Numbers: Properties of Operations				●	●			
Estimating Quantity & Size				•	•			
Decimal Fractions				●	■	●		
Relation of Common & Decimal Fractions				•	■	●		
Properties of Common & Decimal Fractions					●	●		
Percentages					●	●		
Proportionality Concepts					●	●	●	•
Proportionality Problems					●	●	■	■
2-D Geometry: Coordinate Geometry					•	•	●	●
Geometry: Transformations						●	●	●
Negative Numbers, Integers, & Their Properties						•	●	
Number Theory							●	•
Exponents, Roots & Radicals							●	●
Exponents & Orders of Magnitude							•	•
Measurement: Estimation & Errors							•	
Constructions Using Straightedge & Compass							■	•
3-D Geometry							●	■
Geometry: Congruence & Similarity								■
Rational Numbers & Their Properties								•
Patterns, Relations & Functions								•
Proportionality: Slope & Trigonometry								•

• Intended by 4 out of the 6 top-achieving countries
● Intended by all but one of the top-achieving countries (5 out of 6).
■ Intended by all of the top-achieving countries.

SOURCE: Schmidt, Wang, and McKnight (2005). Reprinted with permission from the Taylor & Francis Group.

in schools with multiple tracks, for example. Schools with multiple tracks may in fact perform similarly, on average, but disaggregated results show that while the elite students who are tracked perform at the highest levels, "the kids at the bottom pay the price," performing at lower levels than their counterparts at nontracked schools.

In Schmidt's view, another problem is that too many schools and systems rely on textbooks and such materials as science kits to dictate the curriculum. These resources should support the curriculum, but many textbooks in the United States are crammed with material so they can satisfy every customer: he pointed out that U.S. textbooks are, on average, 800 pages long, in comparison with those in other countries, which are 250-300 pages long. Thus, it is a district's responsibility to reorganize the material to make it coherent and consistent with the standards to which its students are being taught. If all states adopt the Common Core standards, he added, which have been internationally benchmarked and are focused, coherent, and rigorous (see Table 2-4), the result would likely be less tracking and perhaps, eventually, more coherent textbooks. The cur-

TABLE 2-3 Mathematics Curricula of 21 U.S. States

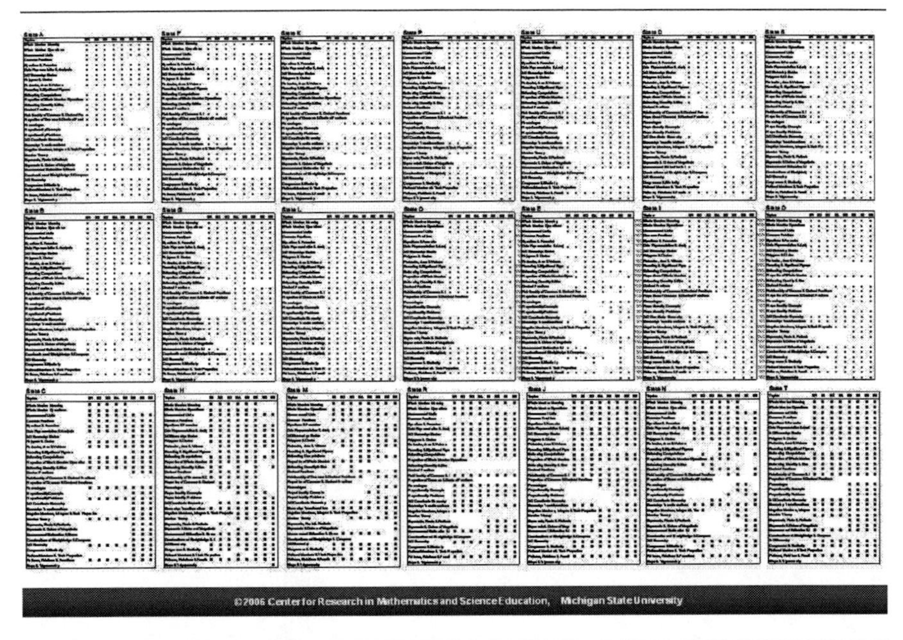

SOURCE: Schmidt (2011). Reprinted with permission.

rent teaching force—another key factor—reflects the deficiencies that have existed in STEM education for some time, Schmidt argued: "We have no standards for teacher preparation and the result is enormous variation."

Schmidt concluded with insights from research he and colleagues are conducting to identify some primary areas of weakness in elementary and middle schools' mathematics instruction to see whether there would be improvements if a more coherent curriculum were implemented. Preliminary results of a randomized trial in 60 districts suggest that the revised curriculum did have a significant effect on learning in specific geometry and algebra topics, such as shape relationships and properties; perimeter, area, and volume; and manipulating expressions. His conclusion from these results is that when students are offered a coherent curriculum, taught by teachers who have been trained to implement it, "they will learn."

USING STATE DATABASES TO IDENTIFY SCHOOL OUTCOMES

This review of school types suggested many factors that may contribute to good outcomes for students. Administrative data collected by states can be used in quantitative analyses that can shed light on the relation-

TABLE 2-4 Common Core Mathematics Standards

Topic	Grade 1	2	3	4	5	6	7	8
Whole Number: Meaning	•	•	•	•	•			
Whole Number: Operations	•	•	•	•	•			
Whole Numbers: Properties of Operations	•	•	•	•	•			
Common Fractions	•	•	•	•	•	•		
Measurement Units	•	•	•	•	•	•	•	•
2-D Geometry: Polygons & Circles	•	•	•	•	•	•	•	•
Data Representation & Analysis	•	•	•	•	•	•	•	•
3-D Geometry	•	•			•	•	•	•
Measurement: Estimation & Errors		•	•					
Number Theory		•		•	•			
2-D Geometry: Basics		•		•	•	•	•	•
Rounding & Significant Figures			•	•	•			
Relation of Common & Decimal Fractions			•	•	•	•		
Estimating Computations			•	•	•		•	•
Measurement: Perimeter, Area & Volume			•	•	•	•	•	•
Equations & Formulas			•	•	•	•	•	•
Decimal Fractions				•	•			
Patterns, Relations & Functions				•	•	•	•	•
Geometry: Transformations				•		•	•	•
Properties of Common & Decimal Fractions					•	•		
Exponents & Orders of Magnitude					•		•	•
2-D Geometry: Coordinate Geometry					•	•	•	•
Exponents, Roots & Radicals					•	•	•	•
Percentages						•	•	
Negative Numbers, Integers, & Their Properties						•	•	
Proportionality Concepts						•	•	•
Proportionality Problems						•	•	•
Rational Numbers & Their Properties						•	•	•
Constructions Using Straightedge & Compass							•	
Systematic Counting							•	
Uncertainty & Probability							•	
Real Numbers, Their Subsets & Properties								•
Geometry: Congruence & Similarity								•
Proportionality: Slope & Trigonometry								•
Validation & Justification								•

SOURCE: Schmidt (2011). Reprinted with permission.

ships between schools' practices and policies and STEM outcomes for students. Michael Hansen described preliminary research he is conducting at the Urban Institute's Center for the Analysis of Longitudinal Data in Education Research with data from Florida and North Carolina. He emphasized that this research is still in progress and that the preliminary exploratory analysis does not support causal inferences.

For Florida, the data available to Hansen included end-of-grade reading and mathematics scores for public school students in grades 3-10 and counts of courses taken in core STEM subjects, advanced STEM, and vocational and technical education, for the school years 2004-2005 through 2008-2009; for North Carolina the same data were available for 2005-2006 through 2008-2009, as well as end-of-course scores.

Looking first at Florida, he noted a few apparent baseline differences among school types (traditional, STEM, and charter or magnet).[10] For example, STEM schools appear to have more new teachers (26 percent, as compared with 21 percent for traditional schools and 23 percent for the charter and magnet schools). STEM schools also are significantly more likely to offer vocational and technical courses (41 percent, compared with 18 and 19 percent for the other types, respectively). At the same time, students in STEM-focused schools take more advanced courses, as might be expected. Hansen was particularly interested in whether expanding access to STEM instruction generally would mean decreased opportunities for high-achieving students, and whether intense focus on STEM for all students would crowd out learning in other subjects. His early findings suggest the possibility that the availability of more advanced courses may tend to push marginal students into lower-track courses. He and his colleagues did not find any negative effects for achievement in reading when more STEM courses were offered.

Hansen also explored whether students in underrepresented minority groups respond differently to variation in STEM opportunities, and, more broadly, whether current approaches are improving STEM outcomes for all students or just those already interested in STEM. His results suggested that when more advanced courses are offered, there is a "pretty strong negative effect" on students who are members of underrepresented minority groups. In other words, "there appears to be a tradeoff" between helping students who are already doing well in STEM subjects and expanding access for all students. The data also suggest benefit from opportunities to conduct research projects in science and from exposure to instruction that was project-based rather than lecture-based. From the preliminary data, Hansen suggested that it appears that teacher characteristics, such as years of experience, are correlated with outcomes for students. From these findings, Hansen concluded that it is important for policy makers to be precise about their goals for STEM education and to focus on specific attributes. But, he added, "we are just beginning to scratch the surface of these databases."

[10] For definitions of these school types, see Hansen (2011).

3

Practices That Support Effective STEM Education

The schools that deliver effective STEM education clearly vary in significant ways, as Adam Gamoran observed. Even within the four primary categories there are marked differences, he noted, and research has not yet provided clear answers as to what makes different approaches work. Thus, it is important to look inside schools at the effects of particular practices and conditions that make them successful. This chapter explores the characteristics of effective science and mathematics instruction, respectively, and then discusses assessment approaches that support STEM instruction.

The STEM fields are interrelated in important ways, and the whole may be greater than the sum of its parts. As one participant noted, mathematics is the language of science, and engineering and technology are both integral to science. Nevertheless, the STEM fields are often treated separately, and science and mathematics are the subject of the most research. Following that research, the workshop and this report focus separately on science and mathematics.

SCIENCE

Richard Duschl described recent approaches focused on treating science in the classroom as a practice, and Okhee Lee discussed ways science education can reach traditionally underserved students.

Teaching Science as a Practice

Richard Duschl noted that the volume of recent reports on the reform of science education demonstrates the attention now focused on the topic. Improving science education has become a "cultural imperative, " essential to the nation's future as a prosperous and democratic state, he said, but he observed that there are several pedagogical challenges. Students are generally not motivated by the economic arguments at the heart of such reports as *Rising Above the Gathering Storm* (National Academy of Sciences, National Academy of Engineering, and Institute of Medicine, 2007) or *Tough Choices or Tough Times* (National Center on Education and the Economy, 2006). Reaching students and helping them to develop as science learners depends instead on instruction that is rich in core knowledge and the practices that are essential to science, such as argument and critique, modeling and representation, and ways in which knowledge is applied. And science is a broad subject, encompassing physics, chemistry, life sciences, and geographic and earth system sciences. Although these subjects all require many of the same tools and technologies, identifying the most important skills and knowledge that students should acquire is not easy.

A recent report offered a vision of science in the context of K-12 education. *Taking Science to School* (National Research Council, 2007) describes science as a social phenomenon, in which a community of peers pursues shared objectives and abides by shared conventions that shape their work, Duschl said. Specifically, science involves practices in which students must learn to engage, such as:

- building and refining theories and models,
- collecting and analyzing data from observations or experiments,
- constructing and critiquing arguments, and
- using specialized ways of talking, writing, and representing phenomena.

Science has evolved, Duschl observed. Not only have technologies become more sophisticated, conceptions of the essential nature of science have also changed. During the first part of the 20th century, Duschl suggested, the focus of science was to test hypotheses and use deductive reasoning to learn from such experiments. Beginning in the 1960s, the focus shifted to the building and revision of theories. In the past two decades, scientists have grown more interested in building and revising models, which are logical representations of the relationships among phenomena that are observed—as opposed to theoretical explanatory frameworks. Regardless of such conceptual distinctions, science has yielded major achievements with both theoretical and practical importance—such as the theory of relativity, the atomic theory of matter, or the germ theory of disease—as well as failures such as crystalline sphere astronomy or theories of spontaneous generation.

Science education does not always fully address the extent of change in science knowledge and practice, and this is one of the reasons why there are ongoing tensions between the way science is conducted and the way science is taught, in Duschl's view. As is noted in *Taking Science to School*, for example, argument is central to science but rare in classrooms. Teaching tends to focus on what students will need to recall, rather than on model-based reasoning about observed phenomena. The norms of the K-12 classroom, where answers are typically provided by teachers and textbooks, are at odds with the way scientists conduct their work, which entails painstakingly building scientific models from accumulating evidence. Curricula and standards that are incoherent and unfocused, and that vary from state to state, work against the logical development of understanding, he said. The demands of the marketplace lead commercial textbook and curriculum developers to focus on stand-alone modules that can be useful in a variety of contexts, rather than on coherent progressions of learning.

Still drawing on *Taking Science to School*, Duschl stressed the importance of teaching the practices of science and engaging students in the kinds of activities in which scientists engage. Doing so means allowing students to design and conduct empirical investigations, linking the investigations to the core knowledge students are developing, working from a curriculum that is linked to meaningful problems, and providing frequent opportunities for students to engage in logical arguments as they learn to build and refine explanations for their observations. Table 3-1 illustrates the relationships among the categories of empirical reasoning students need to develop, scientific practices, and the actions involved in those practices.

Currently, science education does not reflect this approach, Duschl said. *Taking Science to School* found that current curricula and standards:

- contain too many disconnected topics of equal priority,
- use declarative "what we know" language that does not make clear what it means to understand and use knowledge,
- tend to divorce science content from practices, and
- are not sequenced in ways that reflect what is known about the cumulative development of children's scientific understandings.

In contrast, the report advocates a move to the use of learning progressions (National Research Council, 2007). Learning progressions are descriptions of the way students' understanding in a particular discipline

TABLE 3-1 Relationships Among Categories of Empirical Reasoning, Scientific Practices, and Actions

Categories for Empirical Reasoning[a]	Scientific Practices[b]	Verbs[b]
Planning, Designing Data Acquisition	Selection of observation tools and schedule, selection of measurement tools and units of measurement, selection of questions(s), understanding interrelationships among central science concepts, use central science concepts to build and critique arguments	Presents, asks, responds, discusses, revises, expands, challenges, critiques, knows, uses, interprets
Data Collection	Observing systematically, measuring accurately, structuring data, setting standards for quality control, posing controls, forming conventions	Examines, reviews, evaluates, modifies, generates
Evidence (data use)	Use results of measurement and observation, generating evidence, structuring evidence, construct and defend arguments, mastering conceptual understanding	Extends, refines, revises, decides, categorizes
Patterns (modeled evidence)	Presenting evidence; mathematical modeling; evidence-based model building; masters use of mathematical, physical, and computational tools	Represents, evaluates, predicts, discovers, interprets, manipulates, builds, refines, analyzes, models
Explanation	Posing theories, conceptual-based models building, search for core explanation, considering alternatives, understands how evidence and arguments based on evidence are generated, revises predictions and explanations, generates new and productive questions	Builds, refines, represents, interacts

[a]Duschl, and Grandy, (Eds.) (2008).
[b]Michaels, Shouse, and Schweingruber (2008).
SOURCE: Duschl (2011). Reprinted with permission.

develops over time from naïve to sophisticated conceptual understanding (Corcoran, Mosher, and Rogat, 2009). Based on research in neuroscience and other fields that have illuminated many aspects of the way people learn (see National Research Council, 1999, 2001), learning progressions are developed through empirical research on conceptual development related to a specific topic, such as the carbon cycle. They describe goals for the understanding and knowledge that students could be expected to develop by a defined time (e.g., high school graduation), the sorts of misconceptions and naïve understandings students generally begin with, and the intermediate learning steps that lead to the goal of more complete understanding.

Learning progressions are used to coordinate the teaching of knowledge and practices across grades and in the development of assessments that teachers can use to guide students' learning and target their instruction. Standards and curricula that are based in learning progressions support the effective instruction that develops students' understanding of science as a practice. However, the concept of learning progressions is relatively new, Duschl explained, and it is not widely understood. There is a need for more research on students' learning pathways in different domains or subjects, as well as research on ways to use learning progressions effectively in teaching.

Reaching Diverse and Underserved Students

Persistent achievement gaps between student groups are a particular concern in science education because of the increasing economic importance of science and technology, Okhee Lee noted. She described research showing that the gaps in outcomes between "mainstream" students (those who are white, from middle- to high-income families, and are native speakers of standard English) and "nonmainstream" students (students of color, who are from low-income families, and who are learning English as a new language) reflect the different learning opportunities available to these groups (Lee, 2011). Thus, it is critical, she said, to start with the premise that high achievement in science is attainable for most children.

To understand science outcomes for the nonmainstream groups, Lee said, it is important to consider not only standardized test scores, course taking, and school retention or dropout rates, but also these students' opportunities to learn with understanding, to develop an identity as a science learner while also developing their own cultural and linguistic identity, and to develop a sense of agency in their education. Thus, she defined equitable learning environments as those in which (1) the experiences that all students bring from their homes and communities are valued, (2) their cultural and linguistic knowledge is integrated with the disciplinary learning they face at school, and (3) there are sufficient educational resources to support learning. Given these conditions, she explained, nonmainstream students are capable of attaining outcomes comparable to those of their mainstream peers.

Lee described three different perspectives on ways of providing equitable science learning opportunities for nonmainstream students. It is important to consider theoretical approaches, she suggested, because they illuminate underlying mechanisms that apply to different aspects of schooling and different groups, and they also provide a basis for developing strategies to address different challenges.

Cognitive science provides the basis for one approach to understanding how best to promote science learning among students with varying backgrounds. For example, one group of researchers used case studies to explore the ways low-income students from African American, Haitian, and Latino backgrounds in both bilingual and monolingual classrooms engaged in reasoning, problem solving, inquiry, and argument (Rosebery, Warren, and Conant, 1992). The researchers found that the students brought alternative linguistic, conceptual, and imaginative resources to their classrooms but were able to integrate these resources with standard scientific practices. Questioning, argumentation, and innovative uses of everyday words to construct meaning are all practices common in these nonmainstream cultures and also in the practice of science. For example, one aspect of a Haitian oral tradition called *bay odyans* is animated argument about observed phenomena, and teachers can use it to engage their students in scientific discourse in English.

The implication of this cognitive perspective for instruction, Lee observed, is that "when teachers identify and incorporate students' cultural and linguistic experiences as intellectual resources for science learning, they provide opportunities for students to learn to use language, think, and act as members of a science learning community."

Other researchers have explored the ways in which nonmainstream students' cultural traditions may be at odds with Western science as it is practiced and taught, and Lee called this the cross-cultural perspective because it is grounded in the literature on multicultural education. These researchers have examined varying world views and culturally specific patterns of communication and interaction (see, e.g., Snively and Corsiglia, 2001). For example, research on Yup'ik children in Alaska has shown that they learn science-related skills by engaging in activities (such as fishing or navigating by the stars) with adults that build their knowledge over time. In their schools, however, learning is organized around short and frequent lessons in which students are expected to listen, follow directions, and respond quickly to questions verbally and in writing. Though many of the children's scientific ideas may be in harmony with

Western traditions, they need explicit guidance in the "rules of the game," Lee explained. They need to learn to negotiate the boundaries between their own cultural traditions and the expectations of school science, which is described in the literature as "cultural border crossing." When students have this opportunity, she said, they can achieve academically while maintaining their cultural and linguistic identities.

The sociopolitical perspective, which is grounded in critical theory on issues of power, prestige, and privilege, provides another way of thinking about nonmainstream students' and science learning. Researchers working in this tradition, Lee explained, question the value of science as it is currently taught for students who have traditionally been poorly served by the school system (e.g., Calabrese-Barton, 1998; Rodriguez and Berryman, 2002; Seiler, Tobin, and Sokolic, 2001). They suggest that instead of bringing students' world views more in line with science teachers might reconceptualize science to to be more relevant to members of non-mainstream groups.

Studies in this tradition focus on settings in which teachers allow students to take the lead in formulating questions, planning activities, and documenting their explorations. The role of teachers, Lee explained, is to build trust with their students so they are viewed as allies, and the teaching environment is intended to foster the students' cultural identities and sense of agency. Studies of informal science learning, in particular, she noted, suggest that students perform at high levels when they see science as personally meaningful and relevant to their current and future lives, and when they are able to actively engage in it. This research suggests that the mistrust that nonmainstream students bring to the typical classroom is a formidable challenge for their science learning, and that science teachers "must learn to take into account the historical, social, and cultural environments in which their students live," Lee said.

Lee noted that each of these perspectives stresses that finding connections between students' cultural and linguistic experiences and scientific practices is a key to developing equitable learning opportunities. At the same time, they point to a variety of instructional approaches to meeting the learning needs of nonmainstream students. First, teachers need to identify areas in which scientific practices are congruent with students' everyday knowledge and build on them, as the cognitive perspective suggests. Second, teachers need to make the norms and practices of science explicit for students, especially when those norms are at odds with students' experiences, as the cross-cultural perspective suggests. And, third, teachers need to build trusting and caring relationships with their students and engage with them in critical analysis of the purposes of schooling and of science, as the sociopolitical perspective suggests. Perhaps most important, Lee said, is that "a one-size-fits-all instructional approach will surely fail."

MATHEMATICS

The findings from the mathematics literature are similar to those for science, as Jere Confrey and Na'ila Suad Nasir discussed.

Engineering for Effectiveness

Calls to improve schools often focus on the search for "what works," Jere Confrey noted, but a more useful question would be "what works, for whom, and under what conditions?" (Bryk, Gomez, and Grunow, 2011; Means and Penuel, 2005). Many researchers point out that broad scientific principles or guidelines about educational practice are of limited value because the precise conditions in which instruction takes place, the resources available, and other factors have a critical influence on results. Thus, in Confrey's view, a better approach is "engineering for effectiveness," where communities of practitioners and researchers conduct ongoing experiments in a particular context to collect real-time data and use it to tailor improvements, just as engineers might do in an industrial setting.

A number of scholars have suggested variations on this approach, and from their work Confrey has developed an approach to the search for effectiveness that has four elements (Confrey and Maloney, 2011, p. 4):

1. Education must be viewed as a complex system with interlocking parts.
2. Bands and pockets of variability are expected, examined for causes and correlates, and used as sources of insight, rather than adjusted for, suppressed, or controlled.
3. Causal or covarying cycles with feedback and interaction are critical elements of educational systems, in which learning is a fundamental process.
4. Education should be treated as an organizational system that seeks, and is expected, to improve continuously.

Confrey discussed the results from three recent studies of the effectiveness of curricula to illustrate the research approach she advocates. The first was a comparison of the effects of two mathematics curricula for high school students (Grouws et al., 2010). The researchers used a quasi-experimental design in which participants were matched according to their achievement prior to the study to examine and compare the implementation of an integrated mathematics curriculum and a traditional curriculum that treated mathematics subjects (e.g., algebra and geometry) in sequence. The study was conducted in 11 high schools in 6 districts around the country. The populations in the schools ranged from 19 to 23 percent eligible for free or reduced-price lunches. Participating students chose freely between the parallel courses (rather than being tracked) and were evaluated using three outcome measures: two assessments developed for the study—one a test of content common to both curricula, based on a content analysis of the two curricula, and one of reasoning and problem solving. Researchers used multiple data sources—pertaining to factors such as professional development, familiarity with standards, distribution of classroom time among lesson development, noninstruction, practice, and closure—to develop understanding of the relationship of student outcomes to teachers' implementation of the curricula (see Confrey and Maloney, 2011).

The researchers found that, on average, the teachers of the integrated curriculum covered 61 percent of the intended material, and teachers of the single-subject curriculum covered 76 percent. In both courses, teachers also augmented many of the lessons with supplemental material: 28 percent of the integrated course teachers and 33 percent of the single-subject course teachers did so. And what is critical, Confrey said, is not the intended curriculum, but what was actually taught. Thus, to assess this outcome, one needs to look at what was learned.

The preliminary results indicate that the students in the integrated mathematics course made larger gains than the students in the traditional course did, and that having greater opportunity to learn was significantly correlated with high performance. The researchers identified seven factors that influenced the impact of these curricula:

- classroom environment (e.g., the degree of focus on mathematics reasoning and other mathematical thinking);
- fidelity of implementation (e.g., how much of the curriculum was taught);
- use of technology and collaborative learning;
- opportunity to learn;
- teachers' knowledge of the classroom learning environment;
- teachers' experience; and
- teachers' professional development.

Another study compared the results of four different curricula for 1st and 2nd grades (Agodini et al., 2010): "Investigations in Number, Data, and Space," which was categorized as student centered; "Math Expressions," which was categorized as a blend of student and teacher centered; "Saxon Math," which was described as scripted; and the Scott

Foresman-Addison Wesley *Mathematics*, which is a basal textbook. This study also explored the influence of school and teacher characteristics on the implementation of mathematics curricula. The researchers looked at 109 1st grade classes and 70 2nd grade classes in disadvantaged schools that were randomly assigned to one of four curricula.

Confrey noted that the study had an extremely low response rate—just 12 of 473 districts agreed to participate—so it is important to consider whether willingness to participate may be associated with other characteristics that might have an important influence on outcomes. Student results were measured using the Early Childhood Longitudinal Study Measure. The researchers found the highest scores for the students exposed to "Math Expressions" and "Saxon Math." Confrey noted that teachers using "Math Expressions" received more professional development than did teachers using the other curricula, and those teachers also provided more supplements to the curriculum. "Saxon Math" was taught 1 more hour per week than the other curricula. Confrey also noted that both of these curricula were already familiar to the teachers when the study began.

The researchers were careful in defining the elements of adherence to instructional practices consistent with the curriculum developers' intentions, analyzing textbooks and interviewing publishers to make sure what was intended, and then surveying and interviewing teachers to understand what they actually did. Nevertheless, Confrey pointed out, with this study it is difficult to say whether the differences were related to the nature of the curricula or to the specific ways in which they were implemented. Although researchers work hard to maximize the internal validity of such studies, Confrey noted that practitioners will focus not on how results can be generalized but on how the approach might work in their own context.

The third study compared the quality of the implementation of two reform-oriented curricula for grades K-5 in two districts (Stein and Kaufman, 2010). These researchers conducted more than 300 observations (on 3 consecutive days each in the fall and the spring) and were able to cover each of the six grades. They specifically hoped to explore the effects of aspects of high-quality implementation, such as ensuring that instruction places a high-cognitive demand on students, drawing on students' own thinking, and giving the students authority to find solutions.

The researchers supplemented the observations with surveys and interviews. They found that teachers using the *Investigations* curriculum tended to maintain the cognitive demands better, have more emphasis on student thinking, and establish higher classroom norms. In contrast, teachers who used *Everyday Math* reported that frequent shifts of topic in that spiraling curriculum made it more difficult to identify and link the major mathematical concepts they wanted to build.

The study shows that implementation quality cannot be inferred only from content topic analysis but depends also on how the tasks are structured, and appears to relate to the extent of professional development support, facilitated by the district and afforded by the materials, more than to teachers' education, experience, and mathematical knowledge of teaching. The extent to which teachers use the materials to look for "big ideas" correlated with implementation quality across both curricula.

Each of these studies has limitations, Confrey observed, and they do not necessarily support cause-and-effect conclusions. They provide complex results about complex systems, and highlight some critical points. It is very important to be clear about what outcomes the measures are capturing and what factors influence implementation in a particular context before drawing conclusions about a curriculum.

Teachers' capacities and the professional development they receive are critical, Confrey concluded. These are among the elements that define what she called the "instructional core": "If we don't have an effect on the instructional core, we are not going to improve instruction in math and science," she added. Thus, her focus is on helping school systems design the technological capability to gather the information they need and analyze it to support continuous improvement. Noting the important opportunity that states' adoption of the Common Core standards has provided, she closed with a set of specific steps that would constitute a plan for "engineering for effectiveness":

- Construct databases of assessment items linked to the Common Core state standards that can support fair tests of what students are taught.
- Use content analysis to analyze alignment of curricula.
- Build a data system to monitor how curricula are implemented, the ways teachers supplement them, and their reasons for supplementing them.
- Collect data on curricular implementation factors.
- Interconnect the data categories and outcome measures with demographic data for students, classrooms, schools, and districts, and with teacher demographic and survey data.
- Collect teacher demographic and survey data.
- Conduct valid classroom observations and triangulate those data with teacher self-reports.
- Form "networked improvement communities" (see Bryk et al., 2011).
- Define tractable problems.
- Implement continuous improvement models.

Reaching Diverse and Underserved Students

As with science, researchers have explored the mathematics learning of students from nonmainstream groups, including low-income, African American, and Latino students and those with limited English proficiency. Na'ilah Suad Nasir described some of this work (Nasir et al., 2011). She focused on the factors that open up or narrow learning pathways for students, looking particularly at research on the ways schools and society tend to structure lower-quality academic experiences for nonmainstream students. She noted that the majority of the research focuses on students of color and English learners and that the research base is very uneven.

Nasir began by reminding participants of persistent disparities in mathematics achievement. Among 8th graders, for example, just 9 percent of African American and 13 percent of Latino students score at the proficient level, compared with 39 percent of white students. This disparity may be partly explained by the fact that 47 percent of African American students and 49 percent of Latino students complete pre-algebra classes by grade 8, compared with 68 percent of white students. Black and Latino students are also severely underrepresented in honors and advanced placement courses, she added, and there are similar disparities for low-income students. The situation for English language learners is similar, and they are frequently blocked from advanced mathematics tracks because of their lack of English language skills.

Not only are these gaps large, Nasir added, they actually, in many cases, widened in the early 1990s, after a period of narrowing in the 1970s and 1980s. Nasir argued that there are political, economic, and social forces that tend to restrict opportunities for nonmainstream groups and thus to perpetuate the gaps. Schools that serve low-income and minority students tend to have fewer resources, in terms of well-prepared teachers, buildings, supplies, technology, and course offerings. Tracking systems that shuttle black, Latino, and less affluent students into less rigorous educational experiences are both pervasive and rigid. "It is very difficult to jump tracks, Nasir noted, "especially as students move into high school." Discipline systems also tend to disproportionately penalize male black and Latino students, she said, which affects their academic experiences.

Lack of access to a high-quality curriculum and to advanced courses contributes specifically to the achievement gap in mathematics. Classrooms that serve low-income and minority students are much more likely to focus on basics and emphasize instruction that focuses on repetition, practice, and mastering basic arithmetic, Nasir reported from her reading of the research. These conditions have been exacerbated by the recent focus on high-stakes testing, as districts serving nonmainstream students often follow curriculum and instructional practices that have been characterized as teaching to the test in an attempt to increase student scores on state-mandated assessments. As a consequence, the students are provided with fewer opportunities to engage with complex mathematical ideas.

Students with limited English proficiency may be compelled to repeat material they have already learned when they are placed in low-track classes because of their language skills. In addition, because English language learners often switch into their dominant language to engage with higher-level mathematics content, studying only in English before they are fluent may obstruct their access to rich mathematical content. Nasir's own and other research also suggests that students' mathematics learning is also influenced by positive and negative stereotypes and role models, which shape their expectations about who is likely to succeed in mathematics.

The small body of research that compares more and less successful strategies for teaching mathematics to nonmainstream groups highlights two points, Nasir added. The first is that a high-quality curriculum that presents cognitively demanding tasks and builds conceptual understanding and reasoning skills helps students build their skills and become "facile with multiple mathematical representations and multiple solution strategies." Second, classroom practices that foster student-centered discourse and free exploration of mathematical ideas, while addressing multiple kinds of abilities, also help marginalized students learn. "Teachers in successful classrooms find ways to disrupt traditional notions of mathematical competence, such as speed," Nasir explained, "and find ways to assign competence to students who have in the past been unsuccessful in mathematics—for example, by pointing out that particular students ask really good questions."

Additional descriptive research also suggests the importance of approaches in which teachers connect to students' cultural and social backgrounds and focus on building strong relationships with students. This work indicates that when mathematics teachers view equity as a shared mission and work together to "disrupt" the achievement gaps, Nasir said, they are more likely to be successful than when they work on their own.

Nasir used a case study of a California high school she called Railside (to protect students' and faculty members' privacy) that has developed very successful equity practices to illustrate some of the main points she found in the literature. Railside is a large, urban, comprehensive high school with a large nonmainstream student population: it is 80 percent nonwhite, 25 percent of the students are English language learners, and 30 percent qualify for free or reduced-price lunches.

In the late 1990s, Railside abandoned tracking in mathematics. All incoming 9th graders were given the same algebra course, and the school adopted what the staff called a "multi-ability" curriculum, in which

TABLE 3-2 Results for Railside School and Two Comparison Schools (in percentage)

Result or Factor	Railside	Comparison Schools
Students scoring "basic" or better on California Standards Test*	49	41
Seniors in advanced mathematics classes (calculus and precalculus)	41	27
Students who "like mathematics"	74	54
Students interested in mathematics-related careers	39	5

*For information about the California Standards Test, see California Department of Education (2011).
SOURCE: Nasir et al. (2011), data from Boaler and Staples (2008). Reprinted with permission.

instruction in a single classroom could develop a range of mathematical skills. Interviews with Railside teachers showed that they see mathematics teaching as a complex system, Nasir explained: "They work in iterative ways with one another to solve the teaching problems that come up in their classroom. It is not a static approach, but rather a fluid approach [in which] they adapt to the students that are in their classrooms."

Table 3-2 shows data from a study comparing Railside with two other schools in the same Northern California city, indicating that Railside performed well in a variety of outcomes measures (Boaler and Staples, 2008). The study also showed that gender- and race-based achievement gaps were eliminated by students' senior year.

Unfortunately, however, Nasir reported, the Railside mathematics department has recently been under pressure from the district to raise standardized test scores and to use textbooks as the core of their instruction. This pressure has coincided with a district mandate to move from a block schedule, which allowed 90-minute periods, to a schedule with seven 45-minute periods every day and an increase in class sizes (the result of budget cuts in the district). Railside teachers had actually written their own textbook, Nasir explained, and had a system of continuing to rewrite and rework their assignments and activities to make them better, so this was a very frustrating period for them. At the end of the 2009-2010 school year, several of the teacher leaders and former math department heads left Railside because they believed they could no longer sustain their equity practices, Nasir reported: "The remaining teachers feel hopeless about their ability to continue to do the work that they have done together as a department for over 20 years. They say they are just biding time until retirement."

Despite the sobering situation at Railside, Nasir remains optimistic about the potential for schools in which reducing achievement gaps becomes a collective focus for the staff. Yet, she conceded, the research base is not yet sufficient. Studies comparing outcomes for different instructional approaches are needed, as are longitudinal studies that can link classroom practices to equity outcomes. Also important, in her view, will be the development of improved learning measures that can better capture the most important knowledge and skills that students should acquire.

ASSESSMENT

Assessment can have a powerful influence on instruction, for good or ill. James Minstrell described an approach to formative assessment, building on learner thinking (BOLT), that treats assessment and instruction as two facets of a single enterprise (Minstrell, Anderson, and Li, 2011). Formative assessment, he noted, has been defined as "a process used by teachers and students during instruction that provides feedback to adjust ongoing teaching and learning to improve students' achievement of intended instructional outcomes" (Council of Chief State School Officers, 2008). There are a variety of ways to do formative assessment, however, and Minstrell explained that BOLT is based on research on learning and cognition (see, e.g., National Research Council, 1999, 2001). A major finding from this literature is that students bring many kinds of preconceptions to the classroom that affect the way they think about new learning experiences. If instruction does not address the preconceptions that are problematic, students tend to leave the class with those preconceptions intact (National Research Council, 1999). Students also often struggle to transfer what they learn in school to real-world situations, a sign of the limits of their understanding.

Research has shown, however, Minstrell explained, that student performance improves when the teachers and the curriculum purposefully elicit students' thinking about the topic of instruction and address possible misconceptions. Formative assessment is the key to doing so, but there are better and worse ways of using it. Sometimes, Minstrell and his colleagues have found, the focus is on the teacher, on what has been taught, and simply on whether the students "got it" or did not. The results are used to assess the quantity and pace of planned instruction and decide whether the teacher should go on or reteach. More effective, Minstrell said, but less frequently done, is using formative assessment to find out what understandings, including misconceptions or incomplete knowledge, students have, and then to adjust instruction to promote deeper understanding.

The BOLT framework, which has several components, takes this sec-

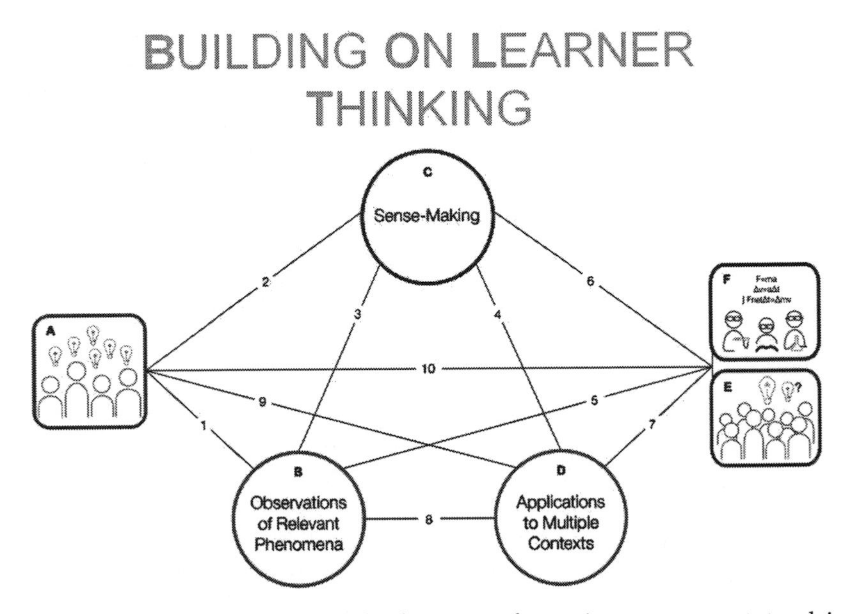

FIGURE 3-1 BOLT: An approach that uses formative assessment to drive instruction.
SOURCE: Minstrell, Anderson, and Li (2011, p. 4). Reprinted with permission.

ond approach: see Figure 3-1. Box A represents the students' ideas prior to instruction on a topic, and boxes E and F represent, respectively, the consensus understanding that a successful class will collectively reach as the students attempt to understand a segment of their curriculum, and the formal ideas of professional scientists. Circle B represents the experiences students have as they learn, which may include observations, tasks, or experiments. Circle C represents another aspect of learning, which Minstrell calls sense-making. Simply doing hands-on activities, he explained, is not sufficient. Students must also mentally process their observations and findings, develop inferences about their meaning, and construct explanations. Finally, circle D represents the many other contexts and representations that promote generalization and the transfer of ideas they produced through the learning experiences, for example, by exploring other hypotheses that may explain the phenomena they have observed.

The lines connecting the circles and boxes represent the ways in which instruction develops the connections among these elements. For example, when a teacher has a clear understanding of the ideas students bring to the topic, he or she can choose or adapt activities and learning opportunities that address those student ideas as well as the learning goals. Each of these connections provides opportunities for "on-the-fly assessment," he explained, using questions such as "how do you know?" or "how can you support that idea?" In contrast to this conceptual approach, however, many typical classroom activities follow a pattern in which the teacher skips students' ideas and presents scientists' ideas, following up with assignments to see whether students have gained procedural and factual knowledge. The activities may be problem sets or lab work in which students are guided to apply the ideas that have been presented and see them in action: these approaches do not typically get students thinking about how the knowledge in question was generated or come up with ideas of their own about how to solve a problem or explain phenomena.

BOLT, instead, focuses on the process of "coming to know" science ideas, Minstrell explained, and the development of the classroom as a "community of science learning." As a class works together to develop consensus in their understanding of the material they are studying, they operate as scientists do. In doing so, they take responsibility for their own learning. The teacher uses formative assessment to identify strengths on which to build and problem areas to address. Diagnostic assessments can be based on "facet clusters," which are a framework for organizing the research on student knowledge and typical misconceptions: see Box 3-1. Facet clusters are derived from standards documents but they also draw

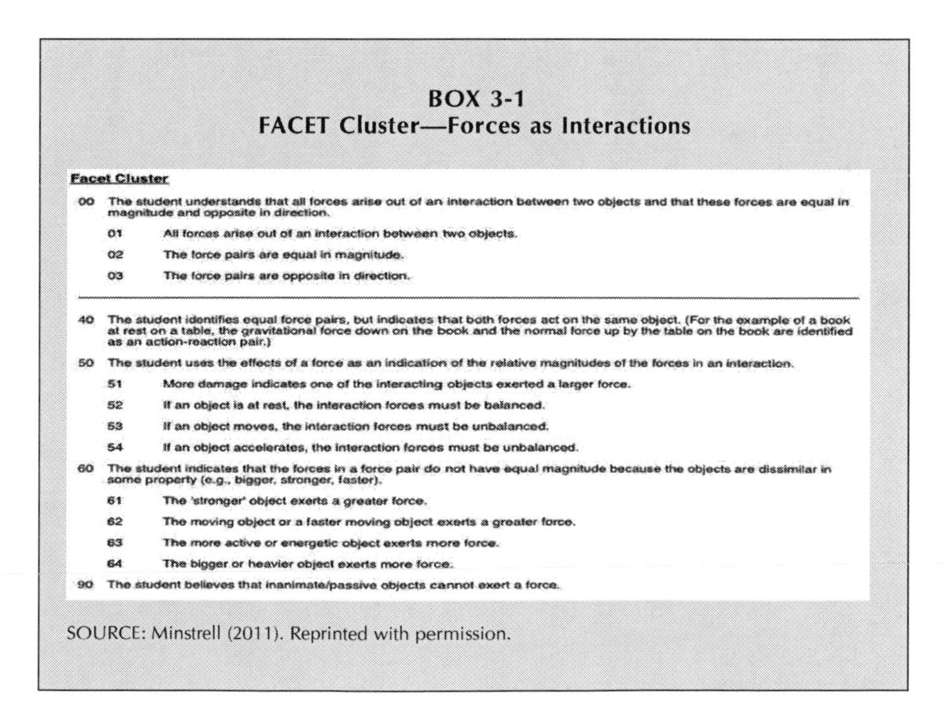

BOX 3-1
FACET Cluster—Forces as Interactions

Facet Cluster

00 The student understands that all forces arise out of an interaction between two objects and that these forces are equal in magnitude and opposite in direction.

- 01 All forces arise out of an interaction between two objects.
- 02 The force pairs are equal in magnitude.
- 03 The force pairs are opposite in direction.

---

40 The student identifies equal force pairs, but indicates that both forces act on the same object. (For the example of a book at rest on a table, the gravitational force down on the book and the normal force up by the table on the book are identified as an action-reaction pair.)

50 The student uses the effects of a force as an indication of the relative magnitudes of the forces in an interaction.

- 51 More damage indicates one of the interacting objects exerted a larger force.
- 52 If an object is at rest, the interaction forces must be balanced.
- 53 If an object moves, the interaction forces must be unbalanced.
- 54 If an object accelerates, the interaction forces must be unbalanced.

60 The student indicates that the forces in a force pair do not have equal magnitude because the objects are dissimilar in some property (e.g., bigger, stronger, faster).

- 61 The 'stronger' object exerts a greater force.
- 62 The moving object or a faster moving object exerts a greater force.
- 63 The more active or energetic object exerts more force.
- 64 The bigger or heavier object exerts more force.

90 The student believes that inanimate/passive objects cannot exert a force.

SOURCE: Minstrell (2011). Reprinted with permission.

on research on problematic student conceptions to describe in detail both explicit learning goals and also difficulties in reasoning and understanding that students are likely to encounter as they progress toward scientifically accurate understanding of the material.

Minstrell acknowledged how difficult it can be for teachers to adopt this approach, noting that one teacher with whom he has worked for many years had explained: "[Y]ou are thinking on your feet constantly. It is draining because you become so intensely involved with your students." Moreover, Minstrell added, "the devil is in the details." Teachers need support not only in how to collect the formative data, but also in how to use it. In response to concerns about how to take the successes the program has had with small groups of teachers to a larger scale, Minstrell added, he and his colleagues have developed a web-based program, called Diagnoser Instructional Tools, which provides learning goals, questions designed to elicit student thinking, developmental lessons, and tools for reporting data to students and teachers students. All the tools are based on the research-based facet clusters.[1]

There is also a need for much more research to support the development of such tools as the facet clusters, Minstrell explained. Much of the existing research on formative assessment has focused on the area of literacy. To reflect the practices of science, research in other kinds of skills will be needed. Moreover, relatively little has been done to explore the ways that formative assessments, such as the BOLT approach, can be used to elicit the cultural influences and perspectives that previous speakers discussed.

[1]See http://www.diagnoser.com/diagnoser/ [July 2011].

4

Conditions That Promote STEM Success in Schools

Whatever approach teachers take in the classroom, their work is affected by numerous factors beyond their direct control. Particularly important are factors that affect teachers' knowledge and skills—their preparation, support for new teachers, and ongoing professional development—and the climate and organization of the schools in which they teach. Suzanne Wilson and Elaine Allensworth addressed these two topics with a focus on teachers and schools, respectively. The final section covers a panel discussion on partnerships between schools and other organizations.

SUPPORTS FOR TEACHERS

There is a significant body of work on questions about teacher preparation, induction into the profession, and continued development for STEM teachers, Suzanne Wilson noted. However, much of it is grounded in a vision of a particular kind of teaching (Wilson, 2011). "That is, practitioners and scholars are interested in teacher support systems that lead teachers to teach in the ways that research and policy suggest they 'should' teach" (Wilson, 2011, p. 2). Wilson reviewed the literature that pertains to STEM teachers, but she noted that researchers have not focused much on subject-specific preparation, induction, or professional development.

Often the developers of a preparation or induction program have a broad goal for changing what is happening in schools. Research about programs with that sort of purpose might have the goal of establishing cause-and-effect relationships or the research might be intended to identify ways to "move the system forward," she observed. Fewer studies examine whether a program prepares teachers who can enhance student learning and engagement.

Research papers in this area also often use teacher preparation or induction as a platform for exploring other issues of interest, she added. For example, because there is considerable interest in the issue of teacher identity in science teacher preparation, the results of many studies focus on claims about teacher identity (Wilson, 2011). Fewer studies directly address questions about what makes particular teacher supports effective.

Nevertheless, as several summaries of the literature on teacher preparation have indicated, a few features are associated with relatively more effective teacher preparation:

- requiring teacher candidates to take more courses in their chosen content area;
- requiring a capstone project (e.g., a portfolio of work done in classrooms or a research paper);
- providing teacher candidates with practical coursework to learn specific practices;
- providing teacher candidates with sufficient opportunities to learn about the curriculum in their local district; and
- providing student teaching experience, carefully overseeing that experience, and ensuring that there is congruence between that experience and later teaching assignments.

The issue of what curriculum teachers are prepared to teach is very significant, Wilson added. Among the approximately 1,200 traditional and more than 140 alternative teacher preparation programs currently in operation, she explained, "most do not know what . . . curriculum . . . their teacher candidates will be teaching."[1] Thus, new teachers must spend time learning what to do with a curriculum they have never seen.

In light of the lack of a core curriculum for teacher preparation, Wilson noted that "some teacher education researchers have begun focusing on core practices" that are key to effective teaching (see Wilson, 2011, p. 5). In particular, she noted, Windschitl et al. (2010) have identified core practices as those that are used frequently with all students, focus on topics that

[1]The characterization of teacher preparation programs as traditional and alternative does not reflect a meaningful distinction, Wilson noted, because these categories overlap markedly in practice. However, research on the differences has been helpful in identifying some of the elements that make teacher preparation effective; see National Research Council (2010) on this point.

are central to the discipline or subject being taught but can apply to different topics and teaching approaches, and can be articulated and taught. These practices can be used by beginning teachers, but they can also be used in increasingly sophisticated ways as teachers gain experience. Core practices should also "play a recognizable role in a larger, coherent system of instruction," Wilson said, that would encompass the content students will learn and the assessments that will be used to track their learning.

The majority of new teachers receive some sort of induction program or support, but there is very little empirical evidence about what aspects of induction make a difference for teachers' effectiveness because very few studies have explored the specific features of these programs. There is some evidence that teachers who participate in an induction program are more likely to stay in the field and to be satisfied with their jobs. This outcome is important because the research shows that the students of teachers who have been in the field for 3-5 years have higher achievement scores than students of newer teachers have, Wilson noted. Some evidence also suggests that coaching is useful and that a match between the coach and the subject matter being taught makes a difference.

The literature on professional development, in Wilson's view, "still consists largely of a nominated list of best practices," though there is some promising research under way. The best practices include the following:

- focusing on developing teachers' knowledge and capacity to teach specific subject matter;
- addressing problems and issues that teachers experience in their classrooms;
- structuring the program around concrete tasks in which teachers teach, assess their students, observe them, and reflect on their practice; and
- allowing sufficient time for teachers to engage in a teacher development program.

Wilson characterized the current state of teacher preparation, induction, and professional development as "a carnival." She chose this metaphor to capture a reality in which there are excellent programs, terrible programs, and many in between, and in which there are many vendors and many sorts of goals. The system is both incoherent and flat, she suggested, in the sense that, for example, "hardly ever is there an opportunity for a teacher to build on what she has learned from a teacher induction program during a professional development program." The system "isn't even loosely coupled," so teachers and teacher candidates must make do with the programs and supports that are available, however haphazard they may be.

Wilson suggested a few reasons why this nonsystem exists. Many different institutions—universities, school districts, vendors, cultural institutions, and funders—play a part in and influence these programs. They answer to different constituencies and have different purposes. They also serve teacher candidates and teachers who vary across many dimensions. "Participants in various teacher development programs enter with a wildly different array of experiences, knowledge, and skill" (Wilson, 2011, p. 15), making it difficult for those who develop and run professional development programs to plan coherent programs.

Moreover, there is no centralization of structure, requirements, goals, or funding—which could be tools for coordinating policies, practices, and resources to support a sustained focus on professional development. Instead, as Wilson noted, "one program will come in with a textbook. Another will come in with collaboration between a school and university researchers on some sort of curriculum or assessment that they have developed." At the same time, states and districts frequently introduce new mandates that require those who develop and run induction and professional development programs to incorporate new information or material. What is missing, Wilson observed, is the coherence and alignment that would allow the system as a whole to pursue clear-cut goals. In this regard, she noted, "We just add things on . . . we do not collectively say 'this is what we are working on.'"

There are levers for influencing the system, however. For example, state policies can address the structure and content of and funding for teacher preparation and supports, as well as the characteristics of students who enter teacher education programs. District policies may affect teacher assignments and the curricula and texts teachers will use. Universities' policies influence the nature and content of their teacher preparation programs, as well as the potential for cross-university attention to content preparation for teachers. Principals may promote a collaborative culture among teachers and also influence the resources (such as models and mentors) available for teachers.

Wilson also noted that there is currently quite a lot of innovation in terms of teacher preparation (less in induction), but that there is not enough good research. She said she concurred with Jere Confrey, who suggested looking at the kinds of variables that might make a difference. She noted that using measures of the value a teacher adds—such as by using student scores at the end of a teacher support program—provides some data, "but what you have to actually do is design studies that look at all these mediating variables" in order to understand what is really effective. A participant also addressed this point, noting that a U.S. Department of Education innovations and improvement program has required evidence of effectiveness before anything can be funded at scale, but that "most of the innovations that were funded were 20 or 30 years old because it took that long to have the kind of evidence of an effectiveness that would be expected."

The fact that the curriculum for teachers is "flat" is perhaps the most fundamental problem, Wilson said, because "there is no way for a teacher to develop her knowledge of the content and content-based teaching practice over time in increasingly sophisticated ways." As a research base grows that illuminates best practices in a richer way, it should provide a stronger basis for policies that bring coherence to the system. Lists of "best practices" at present are easily misinterpreted. For example, an empirically based finding that professional development is most effective when a significant amount of time is allocated to it is often translated simply into a minimum number of hours, regardless of program quality. Additional research on the particular features that make extensive programs effective could provide insights that might allow others to improve the quality of their programs.

SCHOOL CHARACTERISTICS

A lot of hope has been invested in two strategies for improving student achievement, Elaine Allensworth commented. At a time when the pressure is on public schools to prepare all students for college, despite never before having successfully prepared more than about 30 percent, the hope is that improving the teacher workforce and making curricula more rigorous will provide the change that is needed. However, Allensworth stressed, evidence from research on the organizational structure of Chicago's public schools suggests that "even if you get these things perfect, if you don't consider the context and you focus on these strategies narrowly, you are not going to do a single thing to improve student achievement in our underperforming schools and you may actually make it worse."

One study (Bryk et al., 2009) examined 200 Chicago schools, all of which were performing very poorly in the early 1990s: the researchers wanted to know why half of them improved dramatically and half stayed the same or got worse. They were able to use a wealth of data—including longitudinal survey results, student records and test results, and community and crime data—to compare the two groups of schools. All of the schools were in low-income neighborhoods and served student populations that were 90 percent minority. Figure 4-1 depicts the sharp divergence in the performance of the two groups.

Based on their analysis, the researchers concluded that five organizational supports were crucial for school improvement—and made the difference for the 100 schools that improved so dramatically. They presented these supports in a framework because the supports do not have

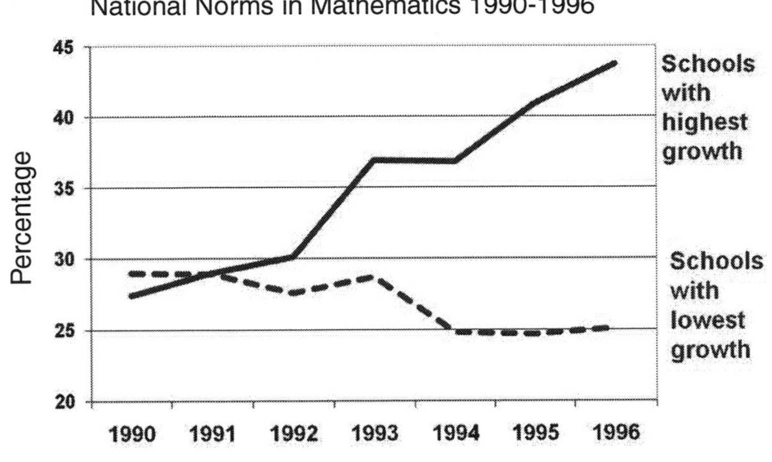

FIGURE 4-1 Mathematics performance of students in Chicago schools with highest and lowest growth, 1990-1996.
SOURCE: Allensworth (2011). Adapted from Bryk et al. (2009).

the same potential benefit in isolation that they have when they function together: see Figure 4-2. For example, principal leadership is necessary, but it must be strategically focused on developing the other four supports, Allensworth explained. The teachers' qualifications were less important than the way in which teachers worked together to take collective responsibility for the school. Similarly, the parents needed not just to participate in school activities, but also to be brought in as partners in their children's education, and community organizations needed to be involved in a way that was aligned with the school's instructional programming. Two other critical elements are a climate that is safe and orderly and supportive to students and an aligned curriculum (that is closely linked to standards) with engaging, student-centered pedagogy.

More specifically, the researchers found that among schools with a well-aligned curriculum and a strong professional community of teachers, 48-57 percent improved substantially in both reading and mathematics. Among schools in which the adults failed to work cooperatively, none improved, and 41-59 percent were stagnant. The real value of these elements, Allensworth explained, lay in their combined strength. Schools that were strong in at least three of the areas were 10 times more likely to improve in reading and mathematics than schools that were weak in

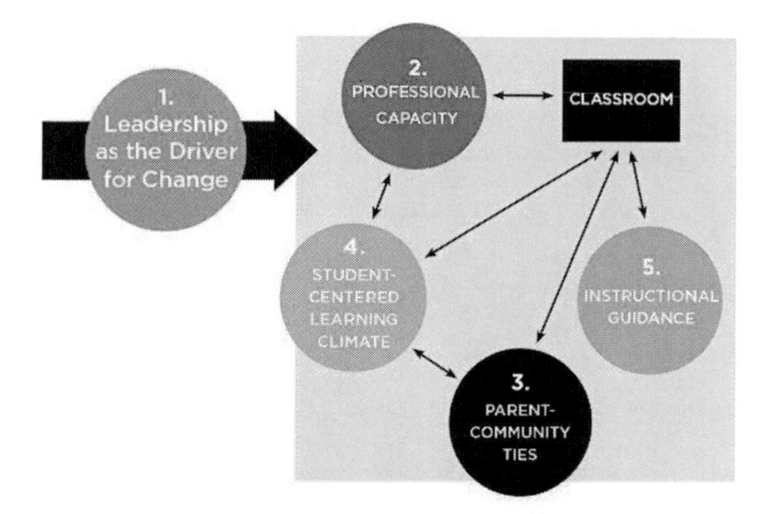

FIGURE 4-2 Five essential supports for school improvement.
SOURCE: Allensworth (2011). Data from Bryk et al. (2009).

three or more. Sustained weakness over time in even one of the elements also appeared to undermine a school's improvement.

The researchers wondered whether these elements are equally important in all types of schools. They divided Chicago's schools into groups on the basis of their racial composition and the economic backgrounds of their students (in Chicago, racial and economic segregation are closely tied). The researchers found that schools serving disadvantaged communities are less likely to show improvements over time: see Figure 4-3. They also found that the most disadvantaged schools are least likely to have the five critical supports. However, if those schools had strong internal supports in all five areas, they were just as likely to improve as advantaged schools that had the supports. The more advantaged schools could better afford to have weaknesses in a few of the elements, but, in general, the essential supports were also more likely to develop in schools in areas where there was strong community cohesion—where people participated in local organizations such as churches and community groups—and where there were lower crime rates.

It may seem obvious that these five elements are important, Allensworth observed, but most improvement strategies are generally much narrower: "Just get the right curriculum. Let's fire all the bad teachers and then hire new ones." Such strategies do not focus on building the organizational capacity of schools. Other studies Allensworth reviewed highlight this point. For example, DeAngelis and Presley (2011) used a

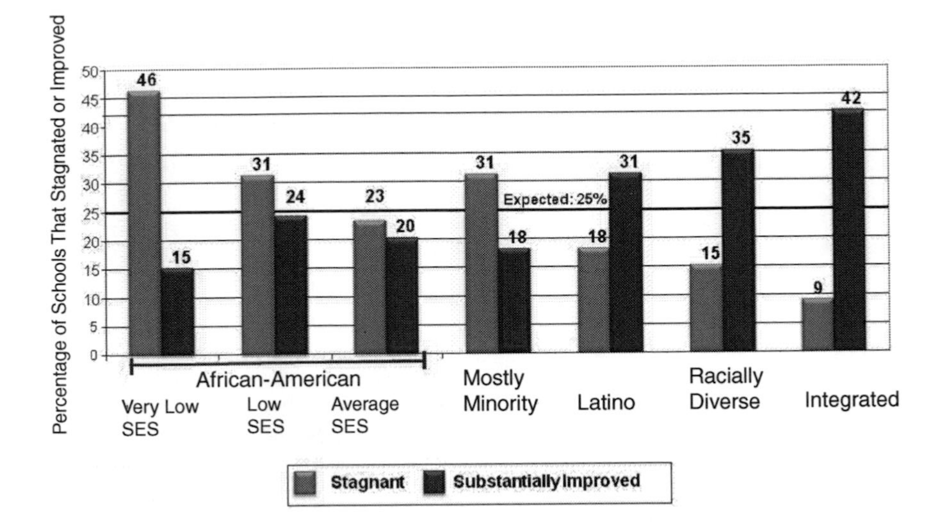

FIGURE 4-3 Comparisons of schools' improvement by characteristics of student populations.
SOURCE: Allensworth (2011). Adapted from Bryk et al. (2009).

wealth of data on Illinois teachers to examine the relationship between their characteristics and student learning. They found that while teacher qualifications were associated with value-added scores in reading and mathematics, organizational structures in the schools actually mattered more. More specifically, schools with highly qualified teachers had much higher learning gains than schools with weaker teachers, but schools that had weak climates—defined by the level of order and safety—did not make gains, even if they had highly qualified teachers. In schools with weak climates, teacher qualifications "made absolutely no difference," Allensworth emphasized. These results suggest that good teachers cannot be effective in schools that lack a supportive climate, and other research shows that teachers leave if they do not believe they can be effective in a school (Allensworth, Ponnisciak, and Mazzeo, 2009).

Allensworth noted that Chicago has also worked hard to reform its curriculum, with the goal of ensuring that all students will take more rigorous classes and there will be more equity among them. She observed that "de-tracking" students was identified by a number of other presenters as a critical improvement tool. In Chicago, however, the result has been a decline in achievement.

Chicago began in 1997 to require all of its students to take a college preparatory curriculum and eliminated all of its remedial classes. There was no concomitant rise in test scores. Instead, attendance and graduation rates declined, more students failed courses, and fewer students went on to 4-year colleges (Allensworth et al., 2009; Montgomery and Allensworth, 2010). Thinking that the curriculum had not been implemented well, the district intensified its focus on aligning the curriculum with professional development and providing teachers with coaching. But increased time on mathematics and improved, interactive pedagogy also brought no improvements in test scores or students' grades (Sporte, Correa, and Hart, 2009).

Follow-up research showed that increased rigor does little to increase learning if schools and classrooms are disorderly—indeed, order is a prerequisite to success. However, maintaining order becomes more difficult when demands on students increase because students tend to withdraw when work gets harder, unless support for them increases as well. When order declines, the learning climate for all students is affected. In addition, schools may not have the professional capacity to teach demanding classes to all students.

Two decades of research in Chicago schools show that there are no "magic bullets," Allensworth concluded. Narrow interventions are tools for making improvements, but they should not be ends in themselves, she argued. "School improvement requires systemic work on multiple fronts" to build the five essential supports: school leadership, parent-community ties, strong professional capacity, a student-centered learning climate, and instructional guidance.

Respondent Milbrey McLaughlin highlighted the importance of themes evident in Allensworth's presentation that were also part of other workshop discussions. The descriptions of individual schools and the discussion of practices that support STEM education provided many examples of what teachers and students gain when schools are sites of collaboration and communal learning. For example, Confrey called attention to the value of collaborations among researchers and practitioners to develop improvements for particular educational contexts. Wilson and Schmidt both called attention to the opportunities that are lost when there is insufficient coordination among the elements that influence STEM education, including curriculum for students and for teacher candidates, textbooks, and professional development. At present, McLaughlin noted, Americans not only disagree about what students should be taught, but also lack both a common framework to determine what success means and a common vocabulary with which to investigate and address problems.

Policy makers have a funny way of responding to such ideas, McLaughlin suggested. She noted that a few years ago, when there was a flurry of attention to the importance of teacher communities, one district responded by issuing lists to teachers of colleagues with whom they were assigned to collaborate. This may seem silly, but there are few clear guides as to how to create teacher communities, she added. At the same time, accountability structures exert an opposing force, often seeming to pit teachers against one another in a competitive sense, rather than encouraging them to share data, collaborate about the approaches they have found to be effective, and speak candidly about the challenges they face.

For McLaughlin, these ideas highlight the need for both micro- and macrolevel policies. As an analogy, she cited the Gates Foundation's investment in vaccines. This is a macro strategy she observed, to target diseases on a large scale. But to be effective it has to be implemented on a micro scale: it only works if every individual in a target population is vaccinated, which requires persistent efforts to reach and educate people community by community.

PARTNERSHIPS TO ENHANCE STEM EDUCATION

In a panel discussion of partnerships between schools and external organizations to enhance their capacity to offer quality STEM education and learning experiences, Martin Gartzman, Vanessa Lujan, and Linda Rosen discussed aspects of the education system that can be positively influenced by different sorts of partnerships.

Seeking a Marriage of Interests

Martin Gartzman discussed multidistrict collaborations that provide support for teachers and administrators. He observed that a number of the presentations at the workshop had demonstrated how easily even very well-intentioned and long-standing reforms (such as that at Railside School) can be undermined by changes in policy, personnel, or direction. Partnerships between schools and outside groups are particularly fragile, and in his view what determines their success is not what they tackle (e.g., professional development, curriculum, or afterschool programs), but whether there is confluence between the needs of the district and the interests of the partner. He believes the primary driver should be the needs of the district, although in many cases external partners have preconceived ideas or research goals and are seeking a partner school in which to implement those ideas. Collaborating and managing the partnership requires considerable time and energy from both parties and works best when the interests of both are served.

Gartzman also agreed that innovations that do not address the core instructional program tend to become "feel-good initiatives." As an example, he described a curriculum partnership between Chicago schools and local museums, in which the museums developed curriculum guides for school field trips. However, the district had not articulated a clear vision of its mathematics and science goals for these trips, and the guides consequently did not address core instructional topics. In response, district staff worked with the museums to replace independent activities structured around their exhibits with activities that drew on museum resources in service of curricular objectives.

Using Partnerships to Build Capacity

Vanessa Lujan emphasized that partnerships can influence district and state policy (from the top down) and teacher and district capacity (from the bottom up). Foundation-supported and community-based partnerships (including informal science institutions such as museums), she suggested, that are focused on afterschool programs, curriculum implementation, and professional development can be designed in part to build networks of leaders—superintendents, science coordinators, curriculum leaders, and lead science teachers and out-of-school-educators—who can bring new skills to their work. Districts and schools may encounter policy barriers, however. For example, schools that have been placed in program improvement status because of inadequate test scores may opt out of such opportunities, she explained. Schools move in and out of this status, which makes it difficult for informal science instititutions to build and sustain partnerships within a district. Teacher turnover and layoffs resulting from severe budget problems also undermine team-building and engagement.

Drawing on Different Kinds of Expertise

Linda Rosen described the contribution of Change the Equation, a network of more than 110 CEOs (chief executive officers, of corporations) who "pledge to connect and align their work to transform STEM learning in the United States." The very existence of the organization, she suggested, sends an important message. The participating companies have been interested in and supported STEM education for a long time, but they recognized that their investments "have not brought the return they might have hoped for." Together, the companies are investing more than half a billion dollars annually, as well as allowing release time for their employees to volunteer for STEM programs during working hours.

The organization partners are increasingly aware of the importance of third-party evidence of effectiveness, however, and they have focused on evaluation, Rosen said. They are looking for programs that are not "dependent on a charismatic visionary," but have been demonstrated to be replicable, she explained. Many of their investments are in nonprofit organizations that develop STEM education programs with a track record for effectiveness. They are willing to invest in formal programs that support teachers and students in schools but are particularly drawn to informal education, in part because schools and districts can be very challenging for them to understand and navigate, while they can work with out-of-school partners more easily and see the impact of their work more immediately. When working with schools and districts, they often seek a commitment from the district so that there is a reasonable expectation that the program can be sustained after the partnership ends.

Overview

All three panelists agreed that it is important to find programs that can be scaled up to benefit not just one or two schools but hundreds, but they also noted how difficult that can be in practice. Gartzman reminded the group of earlier discussions of the importance of school context to outcomes. He suggested that the business community may underestimate what is required to achieve the desired outcomes. A participant noted that the focus on informal partnerships and working around district policies was a cause for concern and wondered what it takes to develop successful partnerships within formal K-12 STEM education.

Lujan agreed with Gartzman that listening carefully to districts to understand the challenges that impede their progress is critical. In the context of the Lawrence Hall of Science's BaySci project, she noted, teachers worried that they could not teach science effectively, given the constraints on classroom time because of testing requirements for mathematics and English language arts. BaySci staff worked with the districts and school leaders to help them convey to teachers that they had "permission" to spend time on science and help them reconcile competing demands from the district, the school, and the classroom.

Rosen added that the CEOs had found success in focusing on formal professional development, and Gartzman cited as just one example the Chicago algebra initiative, which was designed to increase the number of students taking algebra by 8th grade. They worked with Chicago-area universities to help increase the number of teachers who had the preparation and credentials to teach algebra: the universities created a 1-year course, which they taught jointly, as well as a credentialing exam.

5

Looking Ahead

The character of existing schools, the effectiveness of current practices in science and mathematics education, and findings from current research were among the main topics of the workshop, but the committee was eager to build on those discussions and consider possibilities for the future. Near the close of the workshop, they asked the presenters and participants to discuss the implications of the presentations and discussions for implementing the next generation of standards and assessments in the STEM disciplines. The closing discussion also covered policy implications, coming developments in STEM education, and promising areas for future research.

IMPLICATIONS FOR STANDARDS AND ASSESSMENTS

Forty-four states have now adopted the "Common Core" standards[1] and many expect their implementation, and the adoption of new assessments aligned with them, to have a powerful influence on K-12 education. But, Steve Schneider pointed out, that idea has been a long time coming. He cited the Smith and O'Day paper (1992) that described the principles of systemic reform, which was a catalyst for a reform effort that engaged states, districts, and schools around the country. The key components of systemic reform were high-quality standards; alignment of curriculum

[1]For a list of the states and information about the initiative, see http://www.corestandards.org/ [July 2011].

with instruction, assessments, and teacher support (both preservice and inservice); and encouraging all school stakeholders to play their part—all issues that are still very current today.

In Schneider's view, a significant challenge to the success of the earlier reform movement was resistance to any kind of federal mandates regarding standards, even though many of the United States' international competitors have had national standards for many years. Now, in part because of federal incentives offered through the Race to the Top initiative,[2] states which together educate 80 percent of the students in the country are adopting new, common standards. It is possible that this change might actually "move the system," he suggested.

Jere Confrey stressed that the new standards will only be successful to the degree that teachers are well prepared to teach to them at each grade level. If this really happens, she believes, the result would be a meaningful improvement in educational equity and outcomes. She also noted that although the standards were written with explicit attention to learning trajectories, the existing research to support that approach is still uneven, so that in practice, for example, in mathematics, the standards reflect "mathematicians' best logical guesses combined with empirically based learning trajectories." It will be very important to increase the empirical base for these going forward, she noted.

She also cautioned that while formative assessment is a powerful and critical tool, the consortia of states that have formed to work on the next generation of assessments have focused almost exclusively on statewide summative assessments. She expects some to incorporate computer-based testing and possibly performance assessment and most to work to assess higher-level thinking skills, but she expressed doubt that there will be the sort of change in psychometric approaches that was highlighted during the workshop discussion of BOLT, for example. The new standards and assessments hold the promise of significant economies of scale that could allow states to explore formative and diagnostic testing and other innovations. "But," she added, "there is nobody really in charge, and nobody at the federal level can take charge because it would start to not look like state standards."

OTHER STEM-RELATED ACTIVITIES

Conceptual Framework for New Science Education Standards

The National Research Council, in collaboration with Achieve, Inc., the American Association for the Advancement of Science, and the National

---

[2]For more information, see http://www2.ed.gov/programs/racetothetop/index.html [July 2011].

Science Teachers Association has developed a conceptual framework for new science education standards (National Research Council, in press), as Tom Keller explained. A draft of the conceptual framework, which was released in July 2010 for public comment, put forth a vision for science education that makes student engagement the highest priority. It articulates cross-cutting concepts (the "big ideas" of science, such as that matter is made up of units called atoms), core disciplinary ideas in the four major domains of science, and scientific and engineering practices. Jennifer Childress explained that Achieve is going to use the frameworks document to develop specific science standards, and she noted that the implementation of the standards across the participating states will present a significant challenge.

Martin Storksdieck shared a few relevant points from a prior National Research Council workshop.[3] States now generally consider several goals that may previously have seemed radical as part of the job. Three such goals are striving for meaningful equity in educational opportunities, focusing on academic rigor for all, and incorporating data into decision making at all levels. However, Storksdieck said, many of the obstacles that have impeded reform in the past remain: lack of capacity and political will to make significant changes and the inherent limitations of some governmental structures are perhaps two of the most prominent ones. Many tradeoffs are necessary in the pursuit of complex changes, he added, so it is important to focus on the incentives that may influence those one wishes to change. These points from the prior workshop are relevant to K-12 STEM education, he said.

CLOSING THOUGHTS

With regard to the broad question of what makes STEM education effective, Adam Gamoran observed that definitive answers are simply not on the horizon in the short term. There is promising research in progress that can provide some help to policy makers and school leaders, and other studies will eventually yield findings about the efficacy of different school models and the different approaches taken under each of the different models. Yet neither the research findings that are available now nor even the findings that will be available when the research now under way is complete will support general conclusions about the efficacy of different school models. There will still be gaps in the knowledge base.

One possible reason for that is the significant diversity in STEM education, even within each of the basic school types. Effective schools appear to share fundamental goals—such as seeking ways to "transcend the tedium"

[3]For more information about the workshop, see http://www7.nationalacademies.org/bose/Large_Scale_Reform_Homepage.html [July 2011]

that is all too often a part of STEM education—but there are many differences among them. It does seem clear, he suggested, that the context in which schools are operating matters. In a practical sense, that context determines the resources that are available to support the school, such as universities, research organizations, or businesses, that can provide direct support and experience for STEM students. And the context influences the policies that shape the school, such as district rules that do or do not allow school leaders and teachers the flexibility they believe they need to be effective.

Teachers matter greatly to schools' outcomes, Gamoran added, particularly their content knowledge.[4] Other discussions highlighted the vital importance of curriculum—particularly curricular focus—as well as a variety of ways of thinking about curriculum and instruction. He mentioned two views: some argue for tight coherence and consistency of the curriculum, while others emphasize the importance of monitoring students' learning as they develop understanding in a particular domain.

The workshop also revealed several areas where more work is needed, Gamoran observed. Much of the discussion of school types focused on high schools, for example, although grades K-8 are also very important. There was more attention to mathematics and science than to engineering and technology education. These are imbalances that reflect the literature, and they may also reflect the emphasis of current accountability policies. The T in STEM has always been easy to overlook, one participant observed, because it is difficult to define. Is it educational technology? Is it technology as a result of engineering? Technology has not been well incorporated into science standards, and although there are separate standards for it, its place has not been clearly established.

Each of these points suggests fruitful areas for further research and analysis, but committee members ended the workshop with an appreciation for the many creative schools, educators, and others who are already hard at work preparing the next generation of STEM students and workers.

[4]Researchers have identified the importance of pedagogical content knowledge, specific knowledge of how to teach the material in a particular field, as very important to teacher effectiveness. See National Research Council (2010) for more on this point.

References

Agodini, R., Harris B., Thomas, M., Murphy, R., and Gallagher, L. (2010). *Achievement Effects of Four Early Elementary School Math Curricula: Findings for First and Second Graders* (NCEE 2011-4001). Washington, DC: U.S. Department of Education, Institute of Education Sciences, National Center for Education Evaluation and Regional Assistance.

Allensworth, E.M. (2011). *Conditions to Support Successful Teaching: School Climate and Organization.* Presentation at the workshop of the Committee on Highly Successful Schools or Programs for K-12 STEM Education, National Research Council, Washington, DC, May 10-12, 2011.

Allensworth, E.M., Ponisciak, S., and Mazzeo, C. (2009). *The Schools Teachers Leave: Teacher Mobility in Chicago Public Schools*. Chicago: Consortium on Chicago School Research.

Allensworth, E.M., Nomi, T., Montgomery, N., and Lee, V.E. (2009). College preparatory curriculum for all: Academic consequences of requiring algebra and English I for ninth graders in Chicago. *Educational Evaluation and Policy Analysis, 31*(4).

Boaler, J., and Staples, M. (2008). Creating mathematical futures through an equitable teaching approach: The case of Railside School. *The Teachers College Record, 110*(3), 608-645.

Bryk, A., Gomez, L., and Grunow, A. (2011). Getting ideas into action: Building networked improvement communities in education. In M. Hallinan (Ed.), *Frontiers in Sociology of Education*. New York: Springer.

Bryk, A.S., Sebring, P.B., Allensworth, E., Luppescu, S., and Easton, J.Q. (2009). *Organizing Schools for Improvement: Lessons from Chicago.* Chicago: University of Chicago Press.

Calabrese-Barton, A. (1998). Teaching science with homeless children: Pedagogy, representation, and identity. *Journal of Research in Science Teaching, 35*(4), 379-394.

California Department of Education. (2011). *CST Released Test Questions.* Available: http://www.cde.ca.gov/ta/tg/sr/css05rtq.asp [July 2011].

Confrey, J., and Maloney, A. (2011). *Engineering [for] Effectiveness in Mathematics Education: Intervention at the Instructional Core in an Era of Common Core Standards.* Paper prepared for the workshop of the Committee on Highly Successful Schools or Programs for K-12 STEM Education, National Research Council, Washington, DC, May 10-12, 2011.

Corcoran, T., Mosher, F.A., and Rogat, A. (2009). *Learning Progressions in Science: An Evidence-Based Approach to Science*. Philadelphia, PA: Consortium for Policy Research in Education.

Council of Chief State School Officers. (2008). *Program: Formative Assessment for Students and Teachers (FAST)*. Available: http://www.ccsso.org/Resources/Programs/Formative_Assessment_for_Students_and_Teachers_(FAST).html [July 2011].

DeAngelis, K.J., and Presley, J.B. (2011). Teacher qualifications and school climate: Examining their interrelationship for school improvement. *Leadership and Policy in School, 10*(1), 84-120.

Duschl, R. (2011). *STEM Education and the Role of Practices*. Presentation at the workshop of the Committee on Highly Successful Schools or Programs for K-12 STEM Education, National Research Council, Washington, DC, May 10-12, 2011.

Duschl, R., and Grandy, R. (Eds.) (2008). *Teaching Scientific Inquiry: Recommendations for Research and Implementation*. Rotterdam, Netherlands: Sense.

Grouws, D., Reys, R., Papick, I., et al. (2010). *COSMIC: Comparing Options in Secondary Mathematics: Investigating Curriculum, 2010*. Available: http://cosmic.missouri.edu/ [June 2011].

Lee, O. (2011). *Effective STEM Education Strategies for Diverse and Underserved Learners*. Paper prepared for the workshop of the Committee on Highly Successful Schools or Programs for K-12 STEM Education, National Research Council, Washington, DC, May 10-12, 2011.

Levesque, K., Laird, J., Hensley, E., Choy, S.P., Cataldi, E.F., and Hudson, L. (2008). *Career and Technical Education in the United States: 1990 to 2005 Statistical Analysis Report*. Washington, DC: U.S. Department of Education, National Center for Education Statistics.

Means, B., and Penuel, W.R. (2005). Scaling up technology-based educational innovations. In C. Dede, J.P. Honan, and L.C. Peters (Eds.), *Scaling Up Technology-Based Educational Innovations*. San Francisco: Jossey-Bass.

Michaels, S., Shouse, A.W., and Schweingruber, H.A. (2008). *Ready, Set, Science! Putting Research to Work in K-8 Science Classrooms*. Board on Science Education, Center for Education. Division of Behavioral and Social Sciences and Education. Washington, DC: The National Academies Press.

Minstrell, J. (2011). *Building on Learner Thinking (BOLT): A Framework for Assessment in Instruction*. Presentation at the workshop of the Committee on Highly Successful Schools or Programs for K-12 STEM Education, National Research Council, Washington, DC, May 10-12, 2011.

Minstrell, J., Anderson, R., and Li, M. (2011). *Building on Learner Thinking: A Framework for Assessment in Instruction*. Paper prepared for the workshop of the Committee on Highly Successful Schools or Programs for K-12 STEM Education, National Research Council, Washington, DC, May 10-12, 2011.

Montgomery, N., and Allensworth, E.M. (2010). *Passing Through Science: The Effects of Raising Graduation Requirements in Science on Course-Taking and Academic Achievement in Chicago*. Chicago: Consortium on Chicago School Research.

Nasir, N.S., Shah, N., Gutierrez, J., Seashore, K., Louis, N., and Baldinger, E. (2011). *Mathematics Learning and Diverse Students*. Paper prepared for the workshop of the Committee on Highly Successful Schools or Programs for K-12 STEM Education, National Research Council, Washington, DC, May 10-12, 2011.

National Academy of Sciences, National Academy of Engineering, and Institute of Medicine. (2007). *Rising Above the Gathering Storm: Energizing and Employing America for a Brighter Economic Future*. Committee on Prospering in the Global Economy of the 21st Century: An Agenda for American Science and Technology. Committee on Science, Engineering, and Public Policy. Washington, DC: The National Academies Press.

National Academy of Sciences, National Academy of Engineering, and Institute of Medicine. (2009). *Rising Above the Gathering Storm Two Years Later: Accelerating Progress Toward a Brighter Economic Future. Summary of a Convocation.* Planning Committee for the Convocation on Rising Above the Gathering Storm: Two Years Later. Committee on Science, Engineering, and Public Policy. T. Arrison, Rapporteur. Washington, DC: The National Academies Press.

National Academy of Sciences, National Academy of Engineering, and Institute of Medicine. (2010). *Rising Above the Gathering Storm, Revisited: Rapidly Approaching Category 5.* Members of the 2005 "Rising Above the Gathering Storm" Committee prepared for the presidents of the National Academy of Sciences, National Academy of Engineering, and Institute of Medicine. Washington, DC: The National Academies Press.

National Center on Education and the Economy. (2006). *Tough Choices or Tough Times: The Report of the New Commission on the Skills of the American Workforce.* Hoboken, NJ: Jossey-Bass.

National Research Council. (1999). *How People Learn: Brain, Mind, Experience, and School.* Committee on Developments in the Science of Learning. J.D. Bransford, A.L. Brown, and R.R. Cocking (Eds.). Washington, DC: National Academy Press.

National Research Council. (2001). *Knowing What Students Know: The Science and Design of Education Assessment.* Committee on the Foundations of Assessment. J.W. Pellegrino, N. Chudowsky, and R. Glaser, Eds. Board on Testing and Assessment, Center for Education, Division of Behavioral and Social Sciences and Education. Washington, DC: National Academy Press.

National Research Council. (2007). *Taking Science to School: Learning and Teaching Science in Grades K-8.* Committee on Science Learning, Kindergarten Through Eighth Grade. R.A. Duschl, H.A. Schweingruber, and A.W. Shouse, Eds. Board on Science Education, Center for Education. Division of Behavioral and Social Sciences and Education. Washington, DC: The National Academies Press.

National Research Council. (2010). *Preparing Teachers: Building Evidence for Sound Policy.* Committee on the Study of Teacher Preparation Programs in the United States, Center for Education. Division of Behavioral and Social Sciences and Education. Washington, DC: The National Academies Press.

National Research Council. (2011). *Successful K-12 STEM Education: Identifying Effective Approaches in Science, Technology, Engineering, and Mathematics.* Committee on Highly Successful Schools or Programs for K-12 STEM Education, Board on Science Education and Board on Testing and Assessment. Washington, DC: The National Academies Press.

National Research Council. (in press). *A Framework for K-12 Science Education: Practices, Crosscutting Concepts, and Core Ideas.* Committee on a Conceptual Framework for New K-12 Science Education Standards. Board on Science Education, Division of Behavioral and Social Sciences and Education. Washington, DC: The National Academies Press.

Rodriguez, A.J., and Berryman, C. (2002). Using sociotransformative constructivism to teach for understanding in diverse classrooms: A beginning teacher's journey. *American Educational Research Journal, 39*(4), 1,017-1,045.

Rosebery, A.S., Warren, B., and Conant, F.R. (1992). Appropriating scientific discourse: Findings from language minority classrooms. *The Journal of the Learning Sciences, 21,* 61-94.

Schenk, T., Jr., Rethwisch, D., and Laanan, F.S. (2009). *Project Lead the Way: Interim Evaluation Report.* Iowa City: Iowa Department of Education. Available: http://educateiowa.gov/index.php?option=com_content&view=article&id=1846&catid=184&Itemid=1430 [July 2011].

Schmidt, W.H. (2011). *STEM Reform: Which Way to Go.* Paper prepared for the workshop of the Committee on Highly Successful Schools or Programs for K-12 STEM Education, National Research Council, Washington, DC, May 10-12, 2011.

Seiler, G., Tobin, K., and Sokolic, J. (2001). Design, technology, and science: Sites for learning, resistance, and social reproduction in urban schools. *Journal of Research in Science Teaching, 38,* 746-767.

Smith, M.S., and O'Day, J.A. (1991). *Putting the Pieces Together: Systemic School Reform.* CPRE Policy Brief, RB-06-4/91. New Brunswick, NJ: Consortium for Policy Research in Education.

Snively, G., and Corsiglia, J. (2001). Discovering indigenous science: Implications for science education. *Science Education, 85*(1), 6-34.

Sporte, S.E., Correa, M., Hart, H.M., and Wechsler, M.E. (2009). *High School Reform in Chicago Public Schools: Instructional Development Systems.* Menlo Park, CA: SRI International.

Stein, M.K., and Kaufman, J.H. (2010). Selecting and supporting the use of mathematics curricula at scale. *American Educational Research Journal, 47*(3), 663-693.

Stone, J.R. (2011). *Delivering STEM Education Through Career and Technical Education Schools and Programs.* Paper prepared for the workshop of the Committee on Highly Successful Schools or Programs for K-12 STEM Education, National Research Council, Washington, DC, May 10-12, 2011.

Stone, J.R. III, Alfeld, C., and Pearson, D. (2008). Rigor and relevance: Testing a model of enhanced math learning in career and technical education. *American Education Research Journal, 45*(3), 767-795.

Subotnik, R.F., Tai, R.H., and Almarode, J. (2011). *Study of the Impact of Selective SMT High Schools: Reflections on Learners Gifted and Motivated in Science and Mathematics.* Paper prepared for the workshop of the Committee on Highly Successful Schools or Programs for K-12 STEM Education, National Research Council, Washington, DC, May 10-12, 2011.

Wilson, S.M. (2011) *Effective STEM Teacher Preparation, Instruction, and Professional Development.* Paper prepared for the workshop of the Committee on Highly Successful Schools or Programs for K-12 STEM Education, National Research Council, Washington, DC, May 10-12, 2011.

Windschitl, M., Thompson, J., Braaten, M., Stroupe, D., Chew, C., and Wright, B. (2010). *The Beginner's Repertoire: A Core Set of Instructional Practices for Teacher Preparation.* Paper presented at the annual meeting of the American Educational Research Association, April, Denver, CO.

Young, V.M. (2011) *Inclusive STEM Schools: Early Promise in Texas and Unanswered Questions.* Paper prepared for the workshop of the Committee on Highly Successful Schools or Programs for K-12 STEM Education, National Research Council, Washington, DC, May 10-12, 2011.

Appendix A

Workshop Agendas

WORKSHOP ON SUCCESSFUL STEM EDUCATION IN K-12 SCHOOLS
MAY 10-12, 2011
20 F CONFERENCE CENTER
20 F STREET, NW
Washington, DC 20001

Workshop Goals

1. Describe four types of K-12 schools that can support successful education in science, technology, engineering, and/or mathematics (STEM):
   a. Elite or selective STEM-focused schools.
   b. Inclusive STEM-focused schools (those with no admissions criteria).
   c. STEM-focused career and technical education schools or programs.
   d. Effective STEM education in comprehensive, non-STEM-focused schools.
2. Draw on existing data and research to determine the effectiveness these school types.
3. Summarize existing research on various elements that constitute and contribute to effective K-12 education in the STEM disciplines and describe how the implementation of these elements can contribute to highly successful STEM schools.

Tuesday, May 10
20 F Conference Center
Conference Room B

CLOSED SESSION

8:00 a.m.

OPEN SESSION

8:30 a.m. **Welcome**
Robert Hauser, National Research Council
Joan Ferrini-Mundy, National Science Foundation
Norman Augustine, Lockheed Martin (ret.)

9:00 a.m. **Workshop Overview and Context**

This section of the workshop will describe how the committee framed the issues related to the study charge.

Speakers: Adam Gamoran (University of Wisconsin–Madison), steering committee chair
Barbara Means (SRI International), steering committee member

9:15 a.m. **Successful Education in the STEM Disciplines: An Examination of Four School Types**

Session Moderator: Max McGee (Illinois Mathematics and Science Academy), steering committee member

This section of the workshop will include presentations on four types of schools. For each school type, the author will describe the range of school models and goals, the range of outcomes the schools seek to influence and evidence of their effectiveness, strengths and weaknesses, and factors that influence their success. A leader from each school type will respond to the research papers.

9:15 a.m. ***Selective STEM Schools***
Presenters: Robert Tai (University of Virginia) and Rena Subotnik (American Psychological Association)

Respondent: Chancellor Todd Roberts (North Carolina School of Science and Mathematics, Durham, North Carolina)

Q&A and Discussion

10:15 a.m. ***Inclusive STEM Schools***
Presenter: Viki Young (SRI International)

Respondent: Principal Darryl Williams (Montgomery Blair High School, Silver Spring, Maryland)

Q&A and Discussion

11:15 a.m. ***Break***

11:30 a.m. ***STEM-Focused Career and Technical Education***
Presenter: James Stone (National Research Center for Career and Technical Education)

Respondent: Jill Siler (Lake Travis High School, Austin, Texas)

Q&A and Discussion

12:30 p.m. ***Continue discussions over lunch***

1:30 p.m. ***Effective STEM Education in Non-STEM Focused Schools***

Presenter: William Schmidt (Michigan State University)

Respondent: Principal Janet Elder (PS #28, Jersey City, New Jersey)

Q&A and Discussion

2:30 p.m. **Using State Databases to Identify Schools Successful in STEM: Florida and North Carolina**

Session Moderator: Julian Betts (University of California, San Diego), steering committee member

This section of the workshop will feature quantitative analyses of student-level data from state administrative databases. The analyses will explore the relationships between school-level inputs and STEM outcomes.

Presenter: Michael Hansen (Urban Institute)

Q&A and Discussion

3:15 p.m. **Break**

3:30 p.m. **Wrap-Up of Day 1, Overview of Day 2**

The committee, speakers, and audience will discuss the following questions:

- What, collectively, does this research tell us about schools that deliver effective education in the STEM disciplines?
- What are the most important findings related to each school type, and why? What are the policy implications of those findings?
- What are the gaps in our knowledge, and what merits additional study?

4:30 p.m. **Adjourn Open Session**

CLOSED SESSION

4:30-8:30 p.m.

Wednesday, May 11
20 F Conference Center
Conference Rooms A and B

CLOSED SESSION

8:00 a.m.

OPEN SESSION

8:30 a.m. **Welcome and Overview**
Adam Gamoran (University of Wisconsin–Madison), steering committee chair

8:45 a.m. **Practices to Support Effective Education in the STEM Disciplines**

Session Moderator: Jerry Gollub (Haverford College), steering committee member

This section of the workshop will synthesize the research on effective practices in the STEM disciplines. Presenters will describe how implementing these practices can help to create highly successful schools and illuminate some challenges associated with implementation.

8:45 a.m. ***Effective Science Instruction***
Presenter: Richard Duschl (Pennsylvania State University)

Presenter: Okhee Lee (University of Miami)

Q&A and Discussion

9:45 a.m. ***Break***

10:00 a.m. ***Effective Mathematics Instruction***
Presenter: Jere Confrey (North Carolina State University)

Presenter: Na'ilah Suad Nasir (University of California, Berkeley)

Q&A and Discussion

11:00 a.m. ***Assessment to Improve Instruction in the STEM Disciplines***
Presenter: James Minstrell (FACET Innovations)

Q&A and Discussion

11:45 a.m. **Continue discussions over lunch**

12:45 p.m. **Conditions to Promote Schools That Are Successful in STEM**
Session Moderator: Jerry Valadez (California State University, Fresno), steering committee member

This section will focus on some vital elements of successful schools.

12:45 p.m. ***Supports for Teachers***
Presenter: Suzanne Wilson (Michigan State University)

Q&A and Discussion

1:30 p.m. ***School Climate/Organization***
Presenter: Elaine Allensworth (Chicago Consortium of School Research)

Respondent: Milbrey McLaughlin (Stanford University), steering committee member

Q&A and Discussion

2:15 p.m. ***Partnerships to Enhance STEM Education: A Panel Discussion***
Panelists:
Martin Gartzman (University of Chicago)
Vanessa Lujan (Lawrence Hall of Science)
Linda Rosen (Change the Equation)

Q&A and Discussion

3:00 p.m. **Break**

3:15 p.m. **Looking Ahead: The Next Generation of Standards and Assessments**

Session Moderator: Steve Schneider (WestEd), steering committee member

Committee members, workshop presenters, and audience members will discuss the implications of the information presented in the workshop for implementing the next generation of standards and assessments in the STEM disciplines.

4:00 p.m. **Bringing It All Together**

The final session will synthesize the major messages from the workshop, including policy implications and areas for future research.

Speakers: Workshop steering committee members
Subra Suresh, National Science Foundation (tentative)

4:30 p.m. **Adjourn Open Session**

CLOSED SESSION

4:30-8:30 p.m.

Thursday, May 12
Keck Center
Room 205
Washington, DC

CLOSED SESSION

8:30 a.m.-12:00 p.m.

Appendix B

Registered Workshop Particpants

Joan Abdallah, American Association for the Advancement of Science
Vance Ablott, Triangle Coalition
Maya Agarwal, Carnegie Corporation/Institute for Advanced Study
Kelley Aitken
Daniel Aladjem, SRI International
Martha Aliaga, American Statistical Association
Sue Allen, National Science Foundation
Elaine Allensworth, Chicago Consortium of School Research
Ruth Anderson, FACET Innovations
Jennifer Annetta, National Oceanic and Atmospheric Administration
Norm Augustine, Lockheed Martin (retired)
Evra Baldinger, Stanford University
Alexandra Beatty, National Research Council
Katherine Bender, National Aeronautics and Space Administration
Barbara Berns, Education Development Center, Inc.
Julian Betts, University of California, San Diego
Sharon Bowers, National Institute of Aerospace
Ted Britton, WestEd
Sarah Brown, National Aeronautics and Space Administration
David Campbell, National Science Foundation
Nicole Cavino, Opportunity Equation
John Cherniavsky, National Science Foundation
Jennifer Childress, Achieve, Inc.
Ralph Cicerone, National Research Council

Julia V. Clark, National Science Foundation
Donna Clem, Maryland State Department of Education
Jere Confrey, North Carolina State University
Patti Curtis, National Center for Technological Literacy, Museum of Science, Boston
Buffy Cushman-Patz, National Science Foundation
Cecelia Daniels, Success for All Foundation
Richard Duschl, Pennsylvania State University
Janice Earle, National Science Foundation
Francis Eberle, National Science Teachers Association
Janet Elder, PS #28, New Jersey
Stuart Elliott, National Research Council
Michael Feder, Office of Science and Technology Policy
Debra Felix, Howard Hughes Medical Institute
Joan Ferrini-Mundy, National Science Foundation
Adam Gamoran, University of Wisconsin–Madison
Brenda Gardunia, National Science Foundation
Martin Gartzman, University of Chicago
Edward Geary, National Science Foundation
Evan Glazer, Thomas Jefferson High School for Science and Technology
Jerry Gollub, Haverford College
Melvin Goodwin, National Oceanic and Atmospheric Administration Office of Exploration and Research
Loryn Green, Cleveland Metropolitan School District
Mark Greenman, National Science Foundation
E. Jean Gubbins, University of Connecticut
John Hall, PA Alliance for STEM Education
Jennifer Hammond, National Oceanic and Atmospheric Administration
James Hamos, National Science Foundation
Michael Hansen, Urban Institute
Robert Hauser, National Research Council
Susan Haynes, National Oceanic and Atmospheric Administration
Jack Hehn, American Institute of Physics
Monica Herk, National Board for Education Sciences
Katie Hill, DC Public Schools
Margaret Hilton, National Research Council
William Hunter, Illinois State University
Tobias Jacoby, DC Public Schools
Leigh Jenkins, U.S. Department of Education
Mel Jones, DC Public Schools
Marlene Kaplan, National Oceanic and Atmospheric Administration
Michael Kaspar, National Education Association

Thomas Keller, National Research Council
Michael Kelley, Sacred Heart School, Washington DC
Dean Kern, National Aeronautics and Space Administration
Keith Kershner, Research for Better Schools
Lindsay Knippenberg, National Oceanic and Atmospheric Administration
Howard Kurtzman, American Psychological Association
Michael Lach, U.S. Department of Education
Christopher Lazzaro, College Board
Okhee Lee, University of Miami, Florida
Min Li, University of Washington
Vanessa Lujan, Lawrence Hall of Science
Laura Lukes, National Science Foundation
Sharon Lynch, George Washington University
Melissa McCartney, *Science* Magazine/American Association for the Advancement of Science
Catherine McCulloch, Education Development Center, Inc.
Max McGee, Illinois Mathematics and Science Academy
Rosalyn McKissick, Mary B. Martin STEM School/Cleveland Metropolitan School District
Milbrey McLaughlin, Stanford University
Barbara Means, SRI International
Hans Meeder, Meeder Consulting Group
Jeff Mervis, *Science* Magazine/American Association for the Advancement of Science
James Minstrell, FACET Innovations
John Moore, National Science Foundation
Jennifer Mullin, WestEd
Darek Newby, House Appropriations Committee
Rebecca Nichols, American Statistical Association
Natalie Nielsen, National Research Council
Frank Niepold, National Oceanic and Atmospheric Administration
Dave Oberbillig, U.S. Department of Energy, Workforce Development
Barbara Olds, National Science Foundation
Elizabeth Parry, North Carolina State University
Leslie Payne, American Society of Civil Engineers
Greg Pearson, National Academy of Engineering
La Tosha Plavnik, Consortium of Social Science Associations
Stephen Pruitt, Achieve, Inc.
Miriam Quintal, Lewis-Burke Associates
Sarah Rand, University of Chicago
Kacy Redd, Association of Public and Land-grant Universities

Elizabeth Reese, National Research Council
Staci Richard, Office of U.S. Senator Lieberman
Derek Riley, Policy Studies Associates
Lawrence Rivitz, Green Street Academy Foundation, Inc.
Todd Roberts, North Carolina School of Science and Math
Roy Romer, College Board
Linda Rosen, Change the Equation
Lisa Rubenstein, University of Connecticut
Terrie Rust, National Science Foundation
Amy Sabarre, DC Public Schools
Doris Santamaria-Makang, Frostburg State University
Karissa Schafer, U.S. Department of Education
William Schmidt, Michigan State University
Steven Schneider, WestEd
Reid Schwebach, George Mason University
John Seelke, University of Maryland, College Park
Carolyn Seugling, U.S. Department of Education
Niral Shah, University of California, Berkeley
Linda Sherman, National Aeronautics and Space Administration
Jill Siler, Lake Travis High School, Texas
Paula Skedsvold, FABBS
Nancy Spillane, National Science Foundation
Peggy Steffen, National Oceanic and Atmospheric Administration, National Ocean Service
James Stone, University of Louisville
Martin Storksdieck, National Research Council
Na'ilah Suad Nasir, University of California, Berkeley
Rena Subotnik, American Psychological Association
Larry Suter, National Science Foundation
Robert Tai, University of Virginia
Colby Tofel-Grehl, University of Virginia
Meg Town
Mike Town, National Science Foundation
Jermelina Tupas, National Institute of Food and Agriculture, U.S. Department of Agriculture
Jerry Valadez, California State University, Fresno
Elizabeth VanderPutten, National Science Foundation
Jo Anne Vasquez, Helios Education Foundation
Rachel Weinstein, U.S. Department of Education
Antoinette Wells, National Aeronautics and Space Administration
Sue Whitsett, National Science Foundation

Brad Wible, *Science* Magazine/American Association for the Advancement of Science
Carl Wieman, Office of Science and Technology Policy
Daryl Williams, Montgomery Blair High School, Maryland
Suzanne Wilson, Michigan State University
Joyce Winterton, National Aeronautics and Space Administration
Ana Kay Yaghoubian, American Association of University Women
Viki Young, SRI International
Karen Zill, Educational media writer/editor (freelance)

Appendix C

Commissioned Papers

Engineering [for] Effectiveness in Mathematics Education: Intervention at the Instructional Core in an Era of Common Core Standards
Jere Confrey and Alan Maloney

Effective STEM Education Strategies for Diverse and Underserved Learners
Okhee Lee

Building on Learner Thinking: A Framework for Improving Learning and Assessment
Jim Minstrell, Ruth Anderson, and Min Li

STEM Reform: Which Way to Go
William Schmidt

Delivering STEM Education Through Career and Technical Education Schools and Programs
James Stone

Mathematics Learning and Diverse Students
Na'ilah Suad Nasir, Niral Shah, Jose Gutierrez, Nicole Louie, Kim Seashore, and Evra Baldinger

Study of the Impact of Specialized Science High Schools
Rena Subtonik and Robert Tai

Effective STEM Teacher Preparation, Induction, and Professional Development
Suzanne Wilson

Inclusive STEM Schools: Early Promise in Texas and Unanswered Questions
Viki Young

Appendix D

Biographical Sketches of Committee Members

Adam Gamoran (*Chair*) is the John D. MacArthur professor of sociology and educational policy studies and the director of the Wisconsin Center for Education Research at the University of Wisconsin–Madison. His research focuses on inequality in education and school reform. Two of his current studies are large-scale randomized trials, one on the impact of professional development to improve teaching and learning in elementary and one on the impact of a parent involvement program to promote family-school social capital and student success. He also directs an interdisciplinary training program that prepares social science doctoral students to conduct rigorous research on issues of education policy and practice. He is a member of the National Academy of Education, and he currently chairs the congressionally mandated Independent Advisory Panel of the National Assessment of Career and Technical Education for the U.S. Department of Education. He holds a Ph.D. in education from the University of Chicago.

Julian Betts is professor of economics at the University of California, San Diego (UCSD), research associate at the National Bureau of Economic Research, and an adjunct fellow at the Public Policy Institute of California. His research focuses on the economic analysis of education, and he has written extensively on the link between student outcomes and measures of public school spending, including class size, teachers' salaries, and teachers' level of education. His current research includes studies of school choice, San Diego's controversial Blueprint for Student

Success, and California's High School Exit Examination. He also serves on the board of directors of the Preuss School at UCSD, a charter school that admits disadvantaged students from the local area, and on the technical review panel for the longitudinal study of No Child Left Behind. He holds a B.A. in chemistry from McGill University, an M.S. in economics from Oxford University in England, and a Ph.D. in economics from Queen's University in Ontario, Canada.

Jerry P. Gollub is a professor in the natural sciences and a professor of physics at Haverford College, and he is affiliated with the University of Pennsylvania. He has been provost of Haverford College and chair of its Educational Policy Committee. His research is concerned with nonlinear phenomena and fluid dynamics, and he teaches science courses designed for broad audiences on such topics as fluids in nature, predictability in science, and energy options and science policy. He served as chair of the Division of Fluid Dynamics of the American Physical Society and as a member of its executive board, and he is a recipient of the society's fluid dynamics prize and its award for research in undergraduate institutions. He recently served as Leverhulme visiting professor at the University of Cambridge and overseas fellow of Churchill College. He is a member of the National Academy of Sciences, a fellow of the American Academy of Arts and Sciences, and he has served on the board of the National Science Resources Center. He received a Ph.D. in experimental condensed matter physics from Harvard University.

Glenn "Max" McGee is president of the Illinois Mathematics and Science Academy (IMSA). Prior to becoming IMSA's president, he served as superintendent of the Wilmette School District 39 in Wilmette, Illinois. He also previously served as senior research associate at Northern Illinois University Center for Governmental Studies,as state superintendent of education in Illinois and as a principal and teacher in several jurisdictions. His research looked at high-achieving, high-poverty schools that have closed the achievement gap. He is a past chair and current member of the board of the Golden Apple Foundation, and he serves on the board of the Illinois Association of Gifted Children and the Great Books Foundation. He is a member of the Governor's P-20 Council, the Diversifying Higher Education Faculty in Illinois Board, and the Museum of Science and Industry's advisory council. He holds an M.A. and a Ph.D. in educational administration from the University of Chicago.

Milbrey W. McLaughlin is David Jacks professor of education and public policy at Stanford University, director of the John W. Gardner Center for Youth and Their Communities, and codirector of the Center for Research on the Context of Teaching. Her research combines studies of K-12 U.S. education policy and work on the broad question of community-school collaboration to support youth development. Her research on public education focuses on how school teaching is shaped by context issues, such as organizational policy, and the social-cultural conditions of the schools, districts and communities. She is involved with local efforts to engage schools, community organizations and agencies, parents, and faith-based institutions in developing new strategies for promoting youth development. She holds an Ed.M. and a Ph.D. in education and social policy from Harvard University.

Barbara M. Means is codirector of the Center for Technology in Learning at SRI International. She directs SRI's study of science learning in California after-school programs and a national study of how schools are using student data to inform instructional decision making. Her research focuses on ways to foster students' learning of advanced skills through the introduction of technology-supported innovations, and she led the recently completed comprehensive meta-analysis of research on the effectiveness of online learning for the U.S. Department of Education. Other recent work includes a synthesis of cognitive, curriculum, and intervention research on secondary mathematics learning and an examination of high schools with a science, technology, engineering, mathematics focus. She holds an A.B. in psychology from Stanford University and a Ph.D. in educational psychology from the University of California, Berkeley.

Steven A. Schneider is the senior program director of science, technology, engineering, and mathematics at WestEd. He has been the principal investigator of major initiatives on a wide ranage of topics, including cognition and mathematics instruction, assessments and evaluation of student learning, technology and engineering literacy, and an evaluation of California's Statewide Mathematics Implementation Study, He has more than 35 years of experience in science, mathematics, and technology education, including K-12 preservice teacher education, high school science teaching in biology, physics, and oceanography, and professional development. He holds a degree in biology from the University of California, Berkeley, a 6-12 science teaching credential from California State University, San Jose, and a Ph.D. from Stanford University in the design and evaluation of educational programs with an emphasis in science, mathematics, and technology education.

Jerry D. Valadez is director of the Central Valley Science Project at California State University, Fresno. He has 30 years of experience in education as an assistant superintendent, school site administrator, supervisor, curriculum coordinator, program director, instructor and science teacher, and has carried out research pertaining to the preparation of science and mathematics teachers, professional development, teacher quality, student support, English learners, and systemic reform in science, technology, engineering, and mathematics education. He served as president of the National Science Education Leadership Association and on numerous committees and task forces for the National Science Teachers Association. He has also served as special advisor to South Korea with the American Association for the Advancement of Science in developing the first joint international high school summer science academy. He holds an M.A. in educational administration and evaluation from California State University, Fresno, and an Ed.D. in educational leadership from University California, Davis, and California State University, Fresno.